Praise for

THE Naked Brain

"*The Naked Brain* is a fascinating book, taking the reader through the myriad findings of neuroscientists, through brain imaging and the likely function of various brain structures and the effect of certain hormones on behavior. It is written in a manner that makes it accessible to novice and expert alike . . . this book is a must read for anyone who wants to stay one step ahead of the advertisers."
—*The Washington Times*

"An entertaining . . . primer on how the brain functions . . . Restak demonstrates his true talent for applying the scientific to the social, showing how the new brain imaging techniques are being used to read our minds and reshape our society." —*Body & Soul*

"If traditional neuroscience is disturbing, the really creepy stuff is on the horizon, reports neuroscientist Richard Restak . . . he rings alarms." —*The Washington Post Book World*

"Restak avoids both hype and cynicism while providing an enjoyable primer on the future neurosociety." —*Psychology Today*

"Restak compresses a lot of scientific data into accessible language and keeps the narrative conversational . . . an informative and entertaining book." —*Library Journal*

"A good summary of current research, along with some lurid alarm-sounding." —*Kirkus Reviews*

the
naked brain

How the Emerging Neurosociety

Is Changing How We Live,

Work, and Love

RICHARD RESTAK, M.D.

 THREE RIVERS PRESS • NEW YORK

THREE RIVERS PRESS and the Tugboat design are registered trademarks of
Random House, Inc.

Originally published in hardcover in the United States by Harmony Books,
an imprint of the Crown Publishing Group, a division of Random House Inc.,
New York in 2005.

Library of Congress Cataloging-in-Publication Data

Restak, Richard M., 1942–
The naked brain : how the emerging neurosociety is changing how we live,
work, and love / Richard Restak.—1st ed.
1. Brain—Popular works.   I. Title.
QP376. R47285 2006
612.8'2—dc22   2006009440

ISBN 978-1-4000-9809-5

Printed in the United States of America

Design by Lauren Dong

Illustration on page iii by Jeffrey Middleton; illustrations on pages 16 and 71
by Molly Borman; inset illustration on page 16 by Alan Witschonke.

10 9 8 7 6 5 4 3 2 1

First Paperback Edition

To my mother, Alice Hynes Restak

Acknowledgments

John Gabrieli, Jean Decety, Susan T. Fiske, Donald Stuss, Heike Schmolck, Jordan Grafman, Ralph Adolphs, Kenneth Heilman, Jason P. White, Thomas H. Murray, Bettyann Holtzmann Kevles

Contents

the
Naked Brain

Introduction: Welcome to the Neurosociety

During the first half of the twenty-first century our understanding of the human brain will revolutionize how we think about ourselves and our interactions with other people. This revolution is sufficiently powerful that I think it's fair to speak of the emergence of a neurosociety.

As citizens of that neurosociety, we will have to come to terms with brain-based developments such as:

• Tests aimed at revealing to others some of our thoughts and tendencies that, given our choice, we would choose to keep to ourselves
• Brain scan images directed at gauging our suitability for certain jobs
• Tests purporting to explain why we are romantically attracted to some people but repelled by others
• Advertising campaigns that use brain scans to predict the products we are likely to purchase

- Chemical enhancers of the brain designed to turn us into insatiable consumers, even for products and services we don't actually need
- Brain image profiles of us to ascertain which political candidates we are most likely to vote for
- Brain response patterns that reveal the emotions aroused in us by movies and television shows

Developments such as these will become part of everyday life during the first quarter of the twenty-first century, when we will increasingly employ terms and ideas referring to the brain to describe our inner experiences as well as our own and other people's behavior. Indeed, references to the brain are already being used in law and medicine.

Trials of people accused of violent crimes, for example, regularly feature abnormal brain scans as proof of mitigating factors. "There is something wrong with his brain that made him do it" replaces the traditional "There is something wrong with him." Brain scans are also offered to diagnose reading and learning disabilities, as well as to "explain" mental illnesses and suggest ways of treating them.

Incidentally, if you have doubts that such a highly technical topic as neuroscience will exert a powerful social effect ("It's too difficult for the untrained person to understand"), consider the social effects brought about during the twentieth century by three other equally demanding disciplines: computer science, physics, and psychoanalysis. Terms such as *virtual reality*, *fast-forwarding*, *power down*, *relativity*, and *slip of the tongue* are now part of everyday discourse. Brain imaging is currently exerting an equally powerful effect on our ideas about ourselves.

"Nothing has so changed our ideas of the mind within our brains as the pictures of brain scans we now see everywhere—in newspapers and magazines, on television and at the movies," says Bettyann Holtzmann Kevles, Yale University historian and author of *Naked to the Bone: Medical Imaging in the Twentieth Century.*

What's more, our evolving knowledge about the brain has led to the new discipline of social neuroscience: the application of brain science to social interactions. This represents a dramatic break from our usual ways of looking at human behavior.

Traditionally, social and biological approaches to human behavior developed along separate and often antagonistic paths. For instance, when you studied psychology or the social sciences in school you learned how to describe in abstract terms the mutual influences among individuals, groups, societies, and cultures. Anthropologists, economists, linguists, and others provided alternative "explanations." Experts in these diverse fields operated like the fabled blind men palpating an elephant: each expert based his explanations on whatever part of the phenomenon he happened to encounter. Rarely was any reference made to events happening in the brains of the people under study.

Until recently, neuroscientists (brain scientists) were equally restricted in their approach. Over the past two hundred years they have studied the brain in isolation, almost as if they were deconstructing a watch. In order to understand the workings of a watch you don't have to take into account the watch's surroundings (barring extreme conditions of heat or humidity). You simply remove the case—and then tinker with its component parts.

Social neuroscience introduces a whole new dimension based on the recognition that the brain isn't anything like a watch but operates differently depending on social context. Second, an understanding of mind and behavior can be achieved only by merging social and psychological viewpoints with neuroscience: The blind men must start talking to one another, sharing their experiences, questions, and theories.

[handwritten margin note top: How could someone not already know this in the 1970s ?!]

The most fundamental insight provided by ~~social~~ *psychosocial* neuroscience concerns the basically social nature of the brain. First recognition of this occurred in the 1970s with the "discovery" in monkeys that physical contact was even more important than food in determining mother-infant attachment. We have University of Wisconsin primate researcher Harry Harlow to "thank" for this insight.

[handwritten margin note left: this "insight" was gained by extreme cruelty.]

Harlow isolated monkeys from their natural mothers within a few hours after birth and raised them without further contact with their mothers or human substitutes. In the best known of his experiments, baby monkeys interacted with one of two "mother" surrogates constructed of a wooden frame covered by either wire mesh or terry cloth. The baby monkeys preferred the terry-cloth mother and clung to it even if the feeding bottle was attached to the wire mother—a neat demonstration that newborn monkeys have a built-in need for the softness and warmth of maternal care and if deprived of that care will select the next-best substitute.

Subsequent researchers later discovered what was happening in the monkeys' brains during the Harlow experiments. Early loss of physical contact with the mother reduced the number of receptors (binding sites) in parts of the brain to a class of steroid hormones involved in the stress response. Monkeys with decreased receptor binding have a harder time managing stress. And since these receptor-binding changes are permanent, the mother-deprived animal remains susceptible all of its life to stress-related illnesses. In essence, Harlow's discovery demonstrated that an absence of normal social interaction leads to alterations of the brain and, as a consequence, critical changes in behavior that endure over the life span.

Another insight into the social nature of the brain came a decade later when two experts on the effects of drugs on animal behavior discovered an important principle. At the time they were

[handwritten note bottom: There are some extremist "animal liberationists" but that does not justify "experiments" in cruelty, just as valuable information obtained through torture does not justify torturing people.]

conducting a routine experiment on the effects of the stimulant amphetamine on the behavior of male rhesus macaques. Since the drug influenced identical areas in the brains of each monkey, one might expect that one monkey given a drug would respond more or less like another given the same drug. And at first that is pretty much what the researchers observed. But they discovered something quite different when they looked at their experimental results from a fresh perspective.

After taking into account each monkey's dominance position in the social hierarchy, the researchers observed that amphetamine increases dominant behavior in monkeys high in the social hierarchy but increases submissive behavior in monkeys closer to the bottom of that hierarchy. In other words, the researchers noted differences in the monkeys' responses only when they performed both *biological* and *social* analyses. Subsequent research by other scientists confirmed that both biological and social factors must be taken into account in order to gain an accurate understanding about behavior and emotions. This holds not just for animals but for us as well.

For instance, you probably won't have to think very long to confirm from personal experience that threats to social identity produce physical consequences. Most of us get depressed and brood when other people ignore us, or our spouse talks about a divorce, or our boss suggests that perhaps we should look for other work. We become angry when a coworker ridicules us or when we're rejected from a club to which we applied for membership. Indeed, our everyday language is rich with descriptions mixing the biological and the social: "hurt" feelings, "bruised" egos, and "broken" hearts are all metaphors for the painful experiences that can result from social interactions that have gone sour.

Thanks to social neuroscience we can expect to move beyond metaphors over the next several years as we achieve additional insights into the relationship of the brain to our most intimate and personal thoughts.

At the moment, social neuroscience is a fledging discipline similar to aeronautics at the time the Wright brothers made their first flight in Kitty Hawk, North Carolina. At that time who could have predicted the developments in commercial air travel that we now take for granted? A similar situation now exists when it comes to predicting the effects of social neuroscience on our individual and collective lives.

After several years of following research developments in this rapidly evolving field, I've become convinced that precise predictions aren't possible at this early stage of the development. So instead of predictions or a premature attempt at a synthesis of social neuroscience, I've chosen to provide a series of snapshots of how we are beginning to look in new and different ways at our most personal attributes, such as trust, truth telling, love, sympathy, empathy, ethics, competition, dominance, and obedience. Neuroscientists are currently addressing these attributes armed with imaging devices and other research tools capable of measuring brain activity in intervals ranging from hours to milliseconds. Moreover, their findings are influencing marketers, political scientists, and specialists in fields not traditionally associated with the study of the brain.

Some of the chapters that follow will include descriptions of their research; other chapters will suggest thought experiments that you can conduct on your own, testing and observing how your brain responds to social puzzles and conundrums; still others will point out how social neuroscience is influencing the ways we think about ourselves and why we act as we do. In the following ten chapters we will examine what social neuroscience has to tell us about topics such as marketing, economics, even ethics and morality.

In a sentence, the book is about the neurosociety in which we now find ourselves, along with the transformations we can expect in our lives as social neuroscience applications move from the laboratory to the boardroom, the showroom, and the bedroom.

I

The Emergence of the Neurosociety

Brain Imaging: Peering into Bertino's Brain

As a first step in appreciating the impact of social neuroscience, it helps to understand the power of imaging techniques to provide a window into events happening within the brain.

The earliest techniques capable of revealing the brain's inner processing carried a definite risk of injury and sometimes even death. Consequently, they were restricted to patients suffering from various brain diseases. As a result of this emphasis on disease, we presently know more about the functioning of abnormal brains than we know about normal ones. As a neurologist, I'm especially aware of this paradox. Ask me about the brain dysfunctions associated with strokes or autism or even some forms of learning disability and I can explain the difficulties in more detail than you probably want to hear. But ask me how the brain of a genius differs from that of his or her less intellectually gifted counterparts and the explanation isn't going to take long at all.

Not that we can't learn a lot about the normal brain on the basis of studying abnormal brains. Even a study of the diseased brain often provides some helpful insights toward furthering our understanding of the normal brain. My favorite example of this comes from the observations of the late-nineteenth-century Italian experimenter Angelo Mosso. *Was he a doctor or what ??*

In the course of his research Mosso encountered a peasant, Bertino (no last name is recorded), who several years earlier had suffered a head injury severe enough to destroy the bones of the skull covering his frontal lobes (located immediately behind the forehead). The resulting opening, covered only by skin and fibrous tissue, provided Mosso with a window through which he could directly observe the pulsations of Bertino's brain. Similar pulsations can be observed in a newborn baby during the first few weeks of life prior to the growth and fusion of the skull bones. When the baby cries or strains, the pulsations increase; when the baby sleeps, the pulsations subside.

One day when Mosso was observing the pulsations he noticed a distinct increase in their magnitude coincident with the ringing at noon of the local church bells. At this point Mosso, in an act of inspiration, asked the peasant if the ringing of the Angelus reminded him of his obligation to silently recite the Ave Maria. When Bertino responded yes, the pulsations increased again. Intrigued at this sequence, Mosso asked his subject to multiply eight by ten. At the moment Mosso asked the question, the pulsations increased and then quickly decreased. A second increase occurred when Bertino responded with the answer. From this simple but elegant experiment Mosso correctly concluded that blood flow in the brain could provide an indirect measurement of brain function during mental activity.

Inspired by Mosso's findings with Bertino, students of the brain during the early and middle parts of the twentieth century developed more accurate techniques for measuring blood flow and metabolism in the human brain. For instance, dyes and radioactive substances injected into the arteries leading to the brain help pinpoint the relevant structures responsible for vision, movement, and sensation. But one important limitation lessened the usefulness of these probes into the brain's functioning: All of them were intrusive, dangerous, and on occasion fatal. While undergoing one of the tests the subject could suffer a stroke, blindness, even death. Fortunately, that problem is now a thing of the past thanks to the safety of newer techniques, which carry little risk.

Current imaging techniques are often described using a kind of alphabet soup terminology: "The patent's CAT was normal but a contrast-enhanced MRI showed a small SOL in the frontal lobes later confirmed by PET." Such a sentence isn't very helpful to anyone other than a doctor or someone else trained in the use of this off-putting terminology. In place of acronyms and obfuscation, here's a simplified way of thinking about brain imaging.

Basically, imaging techniques are either structural or functional. If you've ever undergone a CAT (shorthand for computerized axial tomography) scan or an MRI (magnetic resonance imaging) scan, the doctor ordering that scan was interested in capturing an image of your brain's *structure*. Perhaps your doctor thought that you might have suffered a stroke or developed a brain tumor. Tumors and strokes can be recognized by the alterations that they bring about in normal brain anatomy. CAT scans and MRI scans provide a picture of those alterations.

Functional imaging, in contrast, depicts what the brain is doing *(=physiology)* over a certain period of time ranging from seconds to minutes. All of the functional imaging techniques (functional MRI, or fMRI, PET scans, and SPECT scans are the most common) are based on a simple principle: Brain activity leads to changes in blood flow (as with Bertino), electrical discharges, and magnetic fields.

[margin note: "doing" something.]

[margin note: Although growing a tumor or blocking a blood vessel are "doing" something — takes longer than seconds or minutes.]

As I'm writing this sentence an fMRI would show increased activity in those areas of my brain associated with thought (especially the frontal areas), vision, and the movement of my fingers across the keyboard of my word processor. An fMRI of your brain would show activation of the visual areas, which process the words on this page, along with the frontal areas, which grasp the meaning of the sentences, and the motor areas, which control the movement of your hand as it reaches up and turns the page.

Ideally, an imaging technique should accurately pinpoint both the structure and the function, the "where" and the "when" of brain activity. On the "where" scale currently available techniques are accurate within millimeters. But the "when" determinations leave a lot to be desired. The temporal resolution of PET (positron-emission tomography) scans is tens of seconds or even minutes. The most technologically advanced fMRI does a bit better, with a resolution on the order of a tenth of a second. But even that is woefully insufficient as a measurement of how rapidly things are happening in the brain. To give you some perspective, consider that an activation originating in the motor neurons of your brain takes only about 150 milliseconds (thousandths of a second) to reach the muscles of your forefinger when you press a doorbell. Or consider that you can accurately identify an object that suddenly enters your field of vision within a few hundred milliseconds.

In short, in order to establish meaningful relationships between our mental lives and events occurring in our brain, it is important to achieve a temporal resolution of milliseconds. But here's the sticky point. The most accurate technique for doing that involves inserting a tiny needle into the brain and then threading it into a single brain cell. While Dr. Strangelove might consider this invasive, potentially risky procedure acceptable in healthy brains, most others would consider it totally unacceptable.

To further appreciate the challenges in depicting brain activity, think back for a moment to the Heisenberg uncertainty principle in quantum physics: You cannot simultaneously determine the position and the velocity of a particle because of the effect created by the act of measurement ("The more precisely the position is determined, the less precisely the momentum is known in this instant, and vice versa," Heisenberg wrote in 1927). Neuroscientists also encounter a kind of uncertainty principle when studying the brain: They have to choose between achieving either an accurate positional fix (within millimeters) or an accurate temporal fix (within fractions of a second). So far no single technique exists that can provide both; only the use of multiple techniques can make possible the desired integration of spatial and temporal information. To further complicate matters, the brain's operation can't be understood by measuring one neuron at a time; instead, we must focus on thousands of neurons firing together to form "circuits."

Everything You'll Need to Know About the Brain

Although a lot will be said about the brain in this book, a detailed knowledge of that incredibly complex structure won't be required. Indeed, all that you'll need is to remain mindful of two useful distinctions.

The first is between controlled and automatic processes. As an example of a controlled process, recall the last time you worked on your income tax or balanced a budget. Your thoughts followed each other in a sequential manner; you remained consciously aware of what you were doing; if requested, you could explain your thought processes to somebody else. In addition, if you continued your efforts long enough you were likely to experience fatigue or boredom.

As an example of an automatic process, think back to the last time you took an immediate dislike to someone you had just met.

Or an occasion when you discerned a hint of condescension in a coworker's voice as she explained a new procedure to you. Or an afternoon when you paused while walking down a street and looked appreciatively as an attractive person passed by. If asked at the time about such occurrences, you would have come up with various explanations to justify your impressions, but these would only have been guesstimates. That's because with automatic processes things *just happen.* You can't really explain why you disliked the new acquaintance while everybody else liked him. Nor why you were the only person who perceived condescension in the coworker. Nor why other people weren't stopping to gawk at that man or woman you found so attractive. Automatic processes, in contrast to controlled processes, involve more than one avenue of thought occurring at a time, don't involve consciousness, aren't accompanied by a sense of effort, and can't easily be explained to anyone else. And since we're not consciously aware of them, automatic processes don't make it onto our mental radar screens.

Most of the things we do involve a necessary balance between controlled and automatic processing. Too much automatic processing and we behave impulsively; too much controlled processing and we become paralyzed by indecision, such as when we mentally rehearse "perfect" responses to every conceivable question we might be asked during the next morning's job interview.

The second distinction is between cognitive and emotional (affective) processes. While cognition is usually defined as "thinking," that doesn't quite capture the elements of what's meant by cognition. Cognition refers to your perceptions of everything that is going on around you, to all of your thoughts and all of the actions that you might take in response to your outer and inner experiences. Here's a definition from a current textbook: "The ability of the central nervous system to attend, identify, and act on complex stimuli." But let's keep it simple: think of cognition as a shorthand term for all of the ways we come to know the world.

Cognition isn't a thing but a process; nobody can hold it in her hand and show it to you, nor can scientific instruments reveal a picture of it. It includes thinking, remembering, daydreaming, mentally calculating—indeed, any mental activity you select can be included under the umbrella term *cognition*.

When you explain to your accountant all of the income tax deductions that you're claiming, you're involved in a cognitive process, essentially repeating aloud the thoughts you previously entertained while working out for yourself the amount of money that you think you owe. Everything is very intellectualized and rational: You talk and he listens. Emotions play little role here. But suppose later that evening you get a call from your accountant informing you that you actually owe additional money to the IRS. At this point, emotional processes—anger and resentment—are likely to displace cognitive processing (at least momentarily). Indeed, you can define an argument as essentially a dialogue in which affect displaces cognition, emotion overpowers reasoning.

While distinguishing between emotions and cognition and between controlled and automatic processes helps in understanding our own and other people's behavior, it still doesn't help us to answer fundamental questions such as: What happens in my brain when I make an impulsive purchase? Or decide that a certain person is trustworthy? Or invest money in a stock after listening to a broker? In order to address such questions from the perspective of the new social neuroscience, look at the diagram of the human brain on page 16. It is the only brain diagram that you will need to understand the topics that we will take up in this book.

The key structures are the medial prefrontal cortex (MPFC), the dorsolateral prefrontal cortex, the temporal lobes, the superior temporal sulcus (STS), the anterior cingulate, the insula, the parietal lobes, the amygdala, the basal ganglia, and the cerebellum.

In order to understand the diagram, keep one organizational

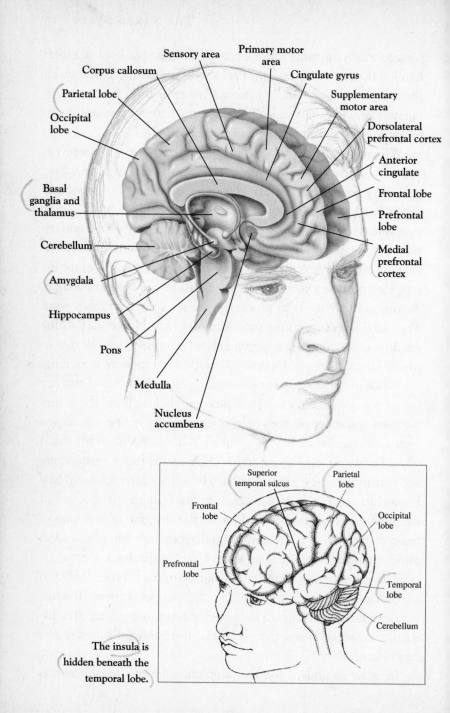

Sensory area

Primary motor area

Corpus callosum

Cingulate gyrus

Parietal lobe

Supplementary motor area

Occipital lobe

Dorsolateral prefrontal cortex

Anterior cingulate

Basal ganglia and thalamus

Frontal lobe

Prefrontal lobe

Cerebellum

Medial prefrontal cortex

Amygdala

Hippocampus

Pons

Medulla

Nucleus accumbens

Superior temporal sulcus

Parietal lobe

Frontal lobe

Occipital lobe

Prefrontal lobe

Temporal lobe

Cerebellum

The insula is hidden beneath the temporal lobe.

principle in mind: automatic and controlled processes occur at different locations in the brain. Those regions that are involved with automatic activity are concentrated toward the back (occipital), top (parietal), and side (temporal) lobes. Controlled processes, in contrast, occur mainly in the front (orbital and prefrontal) areas, with the prefrontal cortex especially important since it integrates information from all other parts of the brain, fashions long- and short-term goals, and directs our overall behavior. Think of the frontal lobes as the CEO of the most complex organization in the world, the human brain. And thanks to the exponential growth of the prefrontal lobes over the past several million years, we are capable of mentally outperforming any other creature on earth.

Since the frontal lobes will be playing a starring role in this book, it's worth spending a few moments to give you a more complete picture of what the frontal lobes do. So let's look first at the kinds of problems that can arise when frontal lobe functioning is compromised.

The Frontal Lobes of Jonathan Meaden

Meet Jonathan Meaden, a sixty-three-year-old man brought by his wife to my office for neurological consultation. After retiring five years ago from a successful career in business, Jonathan devoted himself full time to his lifelong passion: chess. Following a year of concentrated chess study—including tutoring by a local chess master—Jonathan improved his game to the point he could hold his own against highly rated amateur players.

Suddenly things started to go terribly wrong. On several occasions Jonathan became lost while driving to a tournament. In addition, his level of play deteriorated to the point that he placed last in several competitions, losing to players he had consistently defeated in the past. Most of these losses resulted from careless

Dad

mistakes due to failures of concentration. But his problems involved more than just chess and finding his way on the highway.

Always a meticulous and careful investor, Jonathan started
buying questionable oil and gas investments that, according to
his wife, "he would never have had anything to do with before."
Books that formerly interested him now sat unread on the shelves;
if sent to him by relatives or friends, the books sometimes weren't
even taken from their packaging. Even more distressing to his wife
were Jonathan's "memory problems." In addition to forgetting the
names of several of their close friends, Jonathan's memories were
"mixed up," according to his wife; he recalled things out of sequence. But if she raised any questions about accuracy, Jonathan
flew into a rage and began shouting.

On a recent vacation trip to Los Angeles Jonathan left his wife
in the baggage area and rushed ahead to arrange for their rented
car. With the paperwork done, he drove off. Three hours later
Jonathan called his wife from a cell phone and told her to meet
him at a ticket counter in another terminal. When she arrived at
the counter the ticket agent told her that her husband had become impatient after a few minutes of waiting and had gone off to
have lunch. "I was so upset, I was beyond anger. I was utterly flab-
bergasted that he could be so rude and inconsiderate," his wife
said. When she confronted Jonathan about his behavior, he told
her, "You weren't there and I was hungry, so I went and got some
lunch." He seemed unconcerned by the episode and expressed
surprise that his wife was upset about it.

Formerly gregarious, Jonathan now spent most of his time
alone. Always a voracious consumer of newspapers, he now only
glanced at the headlines of the five morning papers delivered to
his home, then turned on the TV and sat for hours "changing
channels with the remote button at a rate that makes me dizzy,"
said his wife. Once a stickler for manners, courtesy, and protocol,
Jonathan now frequently used profane language and recounted

crude jokes to whomever would listen. As another deviation from his usual behavior, he established accounts with several Internet pornography sites. Jonathan's personal habits were also starting to deteriorate. His wife had to remind him to shower, and he frequently went for days at a time without shaving.

At my first meeting with Jonathan at my office I encountered a middle-aged man with a three-day growth of beard and food stains on his necktie. He made little eye contact and during the greater part of our interview sat looking down at the floor. When I asked him why he had come to the office, he denied any problems. "My wife brought me. She has the problems," he responded with a wry smile. This was followed by short, unelaborated responses to my questions. He volunteered no comments of his own.

Routine tests of Jonathan's mental functioning turned up a slightly impaired memory (he could remember only three out of four items four minutes after being asked to remember them for later recall). Most striking was his failure in what's called verbal fluency: rapidly generating as many words as possible within one minute beginning with a particular letter of the alphabet. While the average person can come up with sixteen to twenty words beginning with the letters S or F, Jonathan could only come up with five words apiece. Even more striking, when I asked him to name as many animals as he could in one minute, he named only nine (the normal response is anywhere from twenty to twenty-five).

When asked about a typical day, Jonathan responded that he was spending his time reading chess magazines and engaging in regular chess matches with friends. His wife, seated across the room, indicated by a vigorous shaking of her head that none of this was true. Later I learned from her that Jonathan was very apathetic and didn't get excited by anything, including visits from his daughter and infant granddaughter.

An MRI of Jonathan's brain showed atrophy of the left and

right frontal lobes. A similar pattern of abnormality appeared on a PET scan, with reduced activity in both frontal lobes. The diagnosis was frontal lobe dementia.

The frontal lobes occupy almost a third of the total area of the cerebral hemispheres. Each forms a pyramid with the apex pointing to the front and the base extending toward the back of the brain. Damage anywhere within these triangles results in serious impairments that can affect every aspect of cognitive life. Two principal disorders occur. If the damage involves the upper and lateral surface of the prefrontal cortex, the result is a lack of initiative, loss of creativity, impaired concentration, and a personality marked by blandness and apathy—in short, a great reduction in activity. Damage to the underside of the frontal lobes lying just above the eyes results in too much rather than too little activity: angry outbursts, impulsive acts that may include buying and selling sprees, and tactless, self-destructive behavior toward family, friends, and coworkers. Usually, as with Jonathan, signs of both impairments coexist.

In summary, Jonathan's personality changes included a generalized indifference to other people and events (his self-imposed social isolation), impaired social judgment (the crude jokes and frequent recourse to profanity), diminished emotional responses (the bland, emotionless attitude toward his daughter and granddaughter), faulty practical judgment (the risky oil and gas ventures), and problems with self-control (the unpredictable angry outbursts). Oversimplifying a bit here, Jonathan had great difficulty acknowledging any point of view other than his own. For example, he was genuinely puzzled at his wife's anger at him for needlessly delaying them at the airport. He saw the episode in a different way: Since he had become hungry while waiting for her, he had every right to leave the ticket counter before her arrival and get something to eat.

Jonathan illustrates an extreme example of impairment in empathizing with the thoughts and feelings of other people; he wasn't deliberately choosing to treat his wife and others so callously. Rather, the damage to his frontal lobes robbed him of the ability to place himself in another's shoes and thereby imagine how he would feel under similar circumstances. Neuropsychologists describe people like Jonathan as suffering from a theory of mind (TOM) disorder: a loss of the capacity to infer the internal mental state of another person.

If you've ever interacted with anyone like Jonathan, you will also observe another puzzling aspect of their thinking: They have great difficulty acting in their own best interests. As neurologist Antonio Damasio explains it, damage to the frontal lobes deprives the affected person of a *somatic marker* (defined as the emotional state triggered when thinking of the potential outcomes of one's choices). For example, if you think of not filing a tax return this year, you're likely to experience repetitive feelings of anxiety and dread at the thought of the consequences you'll face if your failure to file is detected. That anxiety and dread serve as somatic markers. In people like Jonathan the somatic markers aren't functioning normally: such people experience no internal disquiet at the prospect of something bad happening to them.

Frontal lobe damage also brings about a strange dissociation between hypothetical and practical decisions. For example, if asked a hypothetical question ("What would you do if you found a stamped addressed envelope on the street?"), the person with frontal lobe damage is likely to give you the correct answer (put the letter in the nearest mailbox). But in a real situation where he actually finds a letter on the street, he is just as likely to open it as to place it in a mailbox. Such a split between saying and doing is customarily considered a hallmark of damage to the frontal lobes. Since people with damage to the frontal lobes can't process anything except from a purely personal perspective, questions of right and wrong are hopelessly muddled.

When Less Is More

Now that we have a general picture of what the frontal lobes contribute to our mental lives, I want to make one final point about brain anatomy and organization: At any given moment our brain is largely under the influence of automatic processes. If that strikes you as a strange arrangement for "rational" creatures such as ourselves, think for a moment about the consequences if everything we did involved controlled process.

When you get up from your chair to walk to the refrigerator, you would have to decide whether to lead with your right or left leg. When you pick up a fork, you would have to stop talking so that you could concentrate on bringing the food to your mouth. And you couldn't even think of driving and talking at the same time. Life would be pretty dull if you had to consciously attend to things such as walking, eating, and driving, wouldn't it? The philosopher Alfred North Whitehead captured how difficult it would be to function effectively if deliberate controlled mental processing were the brain's default mode:

> It is a profoundly erroneous truism . . . that we should cultivate the habit of thinking what we are doing. The precise opposite is the case. Civilization advances by extending the number of operations that we can perform without thinking about them. Operations of thought are like cavalry charges in battle—they are strictly limited in number, they require fresh horses, and must only be made at decisive moments.

A similar suggestion appears in one of my previous books, *Mozart's Brain and the Fighter Pilot*, where I provided a rule for enhancing brain function: Let the brain be the brain. By that I meant, don't pay too much conscious attention (don't "think too hard") when you're trying to learn something new; rather, allow the brain to work on things at its own pace. That's because in many cases we decrease accuracy and efficiency by thinking too hard.

For example, think back to the last time you took a multiple-choice test. If you're like most people, you sometimes found yourself doubting the accuracy of your initial selections on some questions. And the more you thought about those questions, the more uncertain you became about the correct response. According to numerous studies of this experience, your original answer was most likely correct, and additional thinking about the question was likely to lead to errors—one of the reasons students are encouraged after responding to one question to move on to the next question. (Assuming, of course, they're prepared for the test and know the material.) Which Donald "Intuitive" Trump isn't.

Now neuroscientists have an explanation for why we sometimes do better by exerting less rather than greater mental effort. British psychiatrist Paul Fletcher observed via fMRI the brain activity of volunteers instructed to push one of four buttons in response to the highlighting of four boxes on a computer screen. The button they pushed depended on which one of the boxes was highlighted. The subjects were initially asked to press the corresponding button as rapidly as possible without making any errors. Unknown to the subjects, there was a repeating sequence of highlighting (eighteen repetitions of a ten-item sequence). When the volunteers simply relaxed and pushed the buttons, they responded faster and more accurately as time went on, even though they strongly denied noticing any sequence. Later the volunteers were told about the sequence, and asked to try to learn it. But when they actually tried to discern the sequence their responses were slower and less accurate. In other words, those who learned the pattern without consciously trying to do so did better than those who put conscious effort into divining it.

The fMRI results also differed. Among those consciously trying to learn the pattern, a significant increase in activity occurred in the right frontal lobe. And since, as we now know, the frontal lobes are involved in the making of executive-type decisions—decisions that often require the time-consuming weighing of various options—it's no surprise that enlisting the right frontal lobe lengthened response times.

Memory provides another good example of the wisdom of acting spontaneously rather than thinking too much about what response to make. For example, imagine yourself looking very briefly at twenty objects. If I asked you to recall them later, how well do you think you would do? If you're like most adults, you will only be able to come up with about twelve to fifteen. In contrast to this lackluster performance in recollective memory, you will do much better in recognition memory, successfully recognizing up to ten thousand objects when they are re-shown to you after some delay.

Such findings will seem less incredible to you if you consider that most of the things we know exist outside of our conscious awareness; we always know more than we can say. The philosopher Michael Polanyi referred to this as "tacit knowledge." Thanks to tacit knowledge (also called implicit knowledge) we can successfully ride a bike or drive a car even though we aren't able to describe exactly how we do it.

In essence, our brain is organized so that once an activity becomes routine it doesn't require conscious effort but occurs automatically. Indeed, controlled processes come into play only when we encounter something unexpected that forces us to sit back and figure out our response. Or when we have to make a decision about something, especially if we may later have to explain our decision to someone else. Neuroscientists refer to this thinking without conscious effort as the "cognitive unconscious." It plays a pivotal role in social neuroscience and the evolution of the neurosociety.

2

How the Brain Processes Information

The Cognitive Unconscious

In the 1940s movie *Random Harvest* the main character played by Ronald Colman is first introduced as John "Smithy" Smith, a World War I shell shock victim who can't remember anything about his life and identity prior to the war. Through a series of events unlikely in anything but a movie, Smithy meets and marries Paula, an entertainer played by Greer Garson. Months afterward, Smithy is struck by a car while visiting a neighboring town in search of work. The resultant head injury deprives him of his memory of Paula and their recently born child. Now freed from his battlefield-induced amnesia, he reverts to his original identity as Charles Rainier, the scion of a wealthy family, and returns to the magnificent English manor where he grew up.

Throughout the remainder of the movie Rainier becomes uneasy and perplexed whenever he encounters certain people or goes to places such as the pub and tobacco shop Smithy had frequented prior to the car accident.

While it wouldn't be true to claim that Rainier remembers

these past associations—he can't consciously recall them—they nonetheless exert a powerful influence on him. He regularly experiences the feeling that he has been to these places before, yet he can't arouse any conscious memory supporting his feelings. Throughout the movie similar examples of unconscious mental processes provide the main dynamic motivating the actions of Rainier/Smithy.

Although *Random Harvest* is a work of fiction, similar examples of unconscious mental processing can be traced back almost two hundred years. I'm not referring here to the "unconscious" associated with nineteenth-century Vienna, Freud, or the pseudoscience of psychoanalysis. Indeed, the unconscious that I'm referring to has nothing to do with sex, violence, or any of the other factors much beloved by psychoanalysts; rather, it involves how our brain processes information. The operative term now is the *cognitive unconscious*.

One of the first discoverers of the cognitive unconscious was Michael Faraday, the scientist who invented the forerunner of the battery that powers your laptop computer. In 1853 he published a paper in the English periodical *Athenaeum*, "Experimental Investigation of Table Turning." During his research for the paper Faraday had attended several table-turning sessions where, typically, a group of people gathered around a light table and rested their hands on it. After a few minutes—varying from one session to another—the table would rock from side to side, or even turn clockwise or counterclockwise. Since each of the participants denied willfully pressing, pushing, or exerting any force whatsoever on the table, spirits or other supernatural agents were deemed responsible for the table's movement, according to the spiritualist movement popular at the time.

Faraday, however, was convinced that the participants *were* moving the tables but weren't aware that they were doing so. In order to prove his point, he placed force measurement devices between the participants' hands and the table. Sure enough, the source of the table movement turned out to be the participants. A similar explanation also applied to Ouija boards and automatic writing: The participants were unknowingly (Faraday never claimed that they were lying or attempting to deceive anybody) exerting a force absent the usual subjective feeling that accompanies willed action.

Put differently, Faraday had uncovered a split between conscious awareness and intention, on one hand, and the carrying out of a complex series of maneuvers, on the other. This odd arrangement defied any facile attempts to reconcile it with traditional beliefs about consciousness and free will. Granted, we don't always consciously think about everything that we do, but most of us have never had the experience, except in dreams, of thinking that we're not doing something when we actually are. Nor do most of us often feel that an action we carried out actually emanated from some outside agency. According to both our experience and commonsense notions, when we will to do something, we always have the feeling that the action originated with us. But despite our usual feelings of ownership over our own actions, this link between our consciousness and our will isn't quite that simple.

A second example of the cognitive unconscious was described in a 1911 experiment carried out on a woman who had lost her memory as a result of brain damage. Because of the damage she could no longer remember anything or anyone she had encountered only moments earlier.

"She forgot from one minute to the next what she was told, or the events that took place," according to her psychologist, É. Claparède. "When one told her a little story, read to her various items of a newspaper, three minutes later she remembered nothing, not even the fact that someone had read to her." Despite his patient's memory failures, Claparède observed that "with certain questions one could elicit in a reflex fashion some of the details of the forgotten items." What could account for such a puzzling behavior?

To find out, Claparède carried out the following slightly diabolical experiment:

To see whether she would better retain an intense impression involving affectivity [emotions], I stuck her hand with a pin hidden between my fingers. The light pain was as quickly forgotten as indifferent perceptions; a few minutes later she no longer remembered it. But when I again reached out for her hand, she pulled it back in a reflex fashion, not knowing why. When I asked for the reason, she said in a flurry, "Doesn't one have the right to withdraw her hand?" and when I insisted she said, "Is there perhaps a pin hidden in your hand?" To the question, "What makes you suspect me of wanting to stick you?" she would repeat, "That was an idea that went through my mind," or "Sometimes pins are hidden in people's hands." But never would she recognize the idea of sticking as a "memory."

Not only did Claparède's patient not consciously remember the pin, but also she couldn't remember such things as the institution where she had been living for more than six years or the identity of the charge nurse ("With whom do I have the honor of talking?" she would inquire when encountering the nurse). Yet she confidently walked the halls without getting lost, often pausing in her wanderings to ask the nurse whether dinner would soon be ready, or some other domestic question. In short, she unconsciously recognized people, places, and events that she couldn't consciously recall.

Perhaps at this point you're thinking, "All very interesting, but as you mentioned, she suffered from brain damage. Could anything like that happen in a person with a normal brain?" A fair question. Actually, unconscious mental processing can be easily demonstrated in perfectly normal people. Here is a test dating back more than a century that you can use to demonstrate it for your self.

Make up a few index cards printed with a single number or letter. Now seat someone at a sufficient distance away that she can't read the characters on the cards. You will be holding the card at the correct distance when your test subject describes the number or letter on the card as too blurred or dim to discern. ("I can see there is something on the card, but I can't make it out" is the response you will be looking for.) At that point ask her to name what's written on the cards despite her conviction that she can't see the numbers or letters. You'll find that her responses will be correct more often than would be expected on the basis of simple guessing. In the simplest version of the test, use an equal number of cards showing numbers and letters. Ask your subject whether each card taken one at a time has a number or a letter written on it. Most people will perform better than the 50 percent accuracy rate that would be expected on the basis of chance.

This experiment, incidentally, is sufficiently reliable and repeatable that psychology teachers have used it starting in 1898 as a psychology laboratory demonstration. In that year Boris Sides of the Psychology Laboratory at Harvard concluded that this experiment pointed to "the presence within us of a secondary subwaking self that perceives things which the primary waking self is unable to get at."

Over the next hundred years other tests shed additional light on unconscious perception and how it differs from conscious perception. I'll mention just two of them.

Imagine that someone is showing you on a screen a series of clearly visible Chinese ideographs (equivalent to letters or numbers in English). Assume that you don't know any Chinese and can't tell what each ideograph represents. Despite your unfamiliarity with Chinese, you're asked to rate each of the characters as expressing either a "good" concept or a "bad" concept. In such a situation you have no choice but to guess.

Later you will learn that your responses weren't as random as would have been predicted from guessing, but directly related to pictures of human faces expressing happiness (e.g., a smile) or anger (a scowl) that the experimenter, without your knowledge, flashed before your eyes 4 milliseconds preceding each ideograph. Since 4 milliseconds is too short a time for you to consciously perceive the faces, you will have no conscious recollection of having seen them. Nevertheless, the results will indicate that you did perceive them, though without being conscious of doing so. If a smiling face preceded an ideograph, you rated it as representing a "good" concept; conversely, a scowling face led to your identifying the ideograph as representing a "bad" concept.

Now imagine undergoing the same experiment with the pictures flashed slowly enough (e.g., 1,000 milliseconds) for you to consciously register them. You're instructed to ignore the faces and just concentrate on rating the ideographs. Under these conditions the pictures of happy or angry faces don't influence your judgment whether an ideograph is "good" or "bad."

To sum up here, you were able to ignore consciously perceived faces and not let those faces influence your ratings of the ideographs. However, when you were unaware of the rapidly flashed faces, the emotions expressed by the faces influenced your judgment of the ideographs. In other words, unconsciously perceived rather than consciously perceived information exerted a more powerful effect on your emotional reaction to the ideographs. Seems like a somewhat counterintuitive concept, doesn't it?

Not really, according to a test carried out by Phil Merikle, a psychologist at the University of Waterloo, Ontario. In one of Merikle's experiments volunteers sit before a screen while he flashes a series of words. Some of the words come so quickly that they can't be consciously perceived as anything other than a blur; other words come slowly enough for conscious recognition. The critical aspect of the experiment involves a memory test given to the volunteers immediately following the flash of each word on the screen. They are shown the first three letters of the word previously flashed and are requested to complete this word stem with any word *except* that word. For example if the word had been *dough,* then words such as *doubt* or *double* (but not *dough*) are okay to use to complete the stem letters *dou.* Seems simple enough. Yet the volunteers experience great difficulty suppressing the word *dough* following a presentation of that word, which appeared and disappeared so rapidly they hadn't the time to consciously perceive it. In contrast, they had no difficulty suppressing *dough* when the word appeared on the screen slowly enough for conscious recognition.

As Merikle sums it up, "Unconsciously perceived information leads to automatic reactions that cannot be controlled by a perceiver. In contrast, when information is consciously perceived, awareness of the perceived information allows individuals to use this information to guide their actions so that they are able to follow instructions."

Merikle's findings—along with the results of the other experiments discussed above—don't do a lot for one's sense of autonomy. While I'm painfully aware that on occasion I'm influenced by factors not entirely under my control, such as mood and circumstances, I still like to believe that on the whole I'm acting freely. But as the experiments just described make clear, I might just as easily be responding according to the whims of somebody else.

Above and Below the Conscious Threshold

Each of us can bring to mind numerous examples when too much conscious attention mucked up a normally free-flowing process: If we give too much attention to finding the precise word, we lose control of the conversation as a result of our momentary hesitation; if we give in to pressure on the tennis court, we freeze and fail to return the serve; and so on. Similar effects result if we consciously attend to more than one thing at a time. That's because, according to research by psychologist Roy F. Baumeister, acts of conscious self-control in one area interfere with our ability to exert control in another.

Baumeister discovered this operational principle by asking volunteers to simultaneously exert two separate acts of self-control: not eating chocolate cookies that Baumeister placed before them, and not laughing while watching a comedy tape. Baumeister found that even minor acts of conscious self-control (no cookies) interfered with the ability to efficiently carry out the second control task (no laughing). Conscious control, it turns out, is in limited supply, and we pay a price if we attempt to apply it to more than one activity at a time. Like condiments, consciousness is most effective in small pinches; if we try to remain conscious of too many things at a time, we make errors and our efficiency plummets. Baumeister, who has written much on the relationship between consciousness and free choice, estimates that conscious attention plays a causal role in our decisions only about 5 percent or so of the time.

Baumeister's research is in line with Whitehead's insight of a century ago: We advance our civilization—or at least our own best interests—by extending the number of operations that we can perform without thinking about them.

This "separate lives" arrangement within the brain between

conscious and automatic or nonconscious processes represents, according to psychologist John Bargh, "a basic structural feature of the human brain." Since Bargh is one of the nation's experts on nonconscious or automatic influences on social behavior and free will, let's take a moment to examine one of his experiments.

p. 234

Students were exposed to words flashed on the computer screen so rapidly that the students weren't aware of them. Experimenters use this technique, known as *priming*, when they want to activate nonconscious processes. In this particular experiment some of the students were primed with words related to rudeness (e.g., *impolite*, *obnoxious*), others with words denoting politeness (*respect*, *considerate*), and still others with neutral words that have nothing to do with either rudeness or politeness. At this point the class broke up and the students went down the hall, where they encountered a staged situation in which it was possible to act either rudely or politely. Bargh writes, "Although participants showed no awareness of the possible influence of the language test, their subsequent behavior in the staged situation was a function of the type of words presented in that test." In other words, the students who had been primed with words associated with impoliteness were more likely to respond rudely in the staged situation.

In a variation of that test the students were faced with the choice of interrupting or not interrupting a conversation. Sixty-seven percent of students primed with rudeness words interrupted; 38 percent of the students primed with neutral words interrupted; but only 16 percent of those primed with politeness words interrupted.

In another experiment the priming involved either stereotypical words for aging such as *sentimental* and *wrinkle,* or words unrelated to aging stereotypes. Upon leaving the experiment, those primed with the words referring to aging walked more slowly. Memory was also affected if the participants merely responded to a series of questions about elderly people, according to the results of a similar experiment. While these two experiments seem similar, they differ on an important point: In the priming example, the stereotypical words never reached consciousness, while in the question-and-answer experiment the participants retained full conscious access to everything taking place. But despite such a dramatic distinction between the two experiments, each experiment activated within the participants a kind of mental picture (either a conscious or unconscious one) of an aged person. This image then activated the subsequent changes in walking and memory.

Do you find the effects described in the above paragraph surprising—specifically, that it didn't seem to matter whether the experiment involved consciously perceived or unconsciously mediated (subliminal) information? If you're surprised, it's probably because you implicitly assume that the brain distinguishes between what psychologists refer to as supraliminal (above the conscious threshold) and the subliminal (below that threshold). But such an assumption is unwarranted, according to research such as the experiments described above.

"Ninety-nine out of a hundred times the brain does not care at all whether something is primed subliminally or supraliminally," notes social psychologist Ap Dijksterhuis of the University of Amsterdam, the Netherlands. "If the representation of *hostile* or *woman* is activated, certain psychological consequences follow, irrespective of whether activation of the word was the result of subliminal or supraliminal perception. Therefore, if supraliminal activation has a particular effect, subliminal activation of the same thing should have the same effect."

Dreamy States

Additional insights about the brain and the cognitive unconscious evolved from the pioneering research of Wilder Penfield, a philosophical, brooding man with a deep and abiding interest in questions about mind, spirit, and soul. He explored these areas via a series of operating-room experiments on patients undergoing brain surgery for epilepsy. At various point during the operation Penfield recorded his patient's responses as he applied a weak electrical current to their brain. Since, contrary to popular notions, the human brain isn't pain-sensitive, the patients didn't have to be anesthetized but could remain awake and responsive to Penfield's questions.

Penfield discovered that stimulation of areas in the temporal lobes led to the patients describing dreamlike flashbacks to earlier moments in their lives. "It was evident at once that these were not dreams," wrote Penfield. "They were electrical activations of the sequential record of consciousness, a record that had been laid down during the patient's earlier experience. The patient 're-lived' all that he had been aware of in that earlier period of time as in a moving picture 'flashback.' " Since the patients remained fully aware of the unreality of these experiences (many of them used the term "dreamy states"), Penfield referred to them as "experiential illusions." Penfield never claimed, incidentally, that electrode stimulation determined attitudes or behavior: "There is no place in the cerebral cortex where electrical stimulation will cause a patient to believe or decide."

A decade later, Roger Sperry, a tall, gaunt psychobiologist from the California Institute of Technology, made further inroads into what would eventually be called the cognitive unconscious. Before turning his attention to humans, Sperry had already performed research (for which he would eventually win a Nobel Prize) on the nervous systems of a significant number of different animals. Several of these experiments involved the corpus callosum, a tract of nerve fibers connecting one side of the animal's brain with the other.

Cutting the corpus callosum of a cat or a monkey produces a strange creature indeed. Although a "split-brain monkey" (quoting from Sperry here) is "virtually indistinguishable from normal cage mates . . . one finds that each of the divided hemispheres now has its own independent mental sphere or cognitive system—that is, its own independent perceptual, learning, memory, and other mental processes . . . it is as if the animals had two separate brains."

For instance, a split-brain cat with a blindfold over one eye learned to associate a picture of a triangle with the receipt of some liver pâté. But if the blindfold was switched to the other eye, the animal would seem to forget the association. Memory failure wasn't the issue. Rather, the triangle-pâté association disappeared because the linkage between the triangle and the pâté learned by one side of the brain failed to transfer to the other side as a result of the disruption of the corpus callosum linking the animal's two hemispheres.

Would the same principle hold for humans? Since deliberately planned experiments along these lines were obviously impossible in people, Sperry studied patients recovering from brain surgery that involved cutting the corpus callosum (consisting in humans of more than a million connecting fibers). Although nothing would seem to justify such drastic surgery, the reasoning supporting its limited use was actually quite sound. Specifically, seizure discharges in some epileptics reverberate from one hemisphere to the other via the corpus callosum. As a consequence of the resulting unchecked seizures, a productive person can be turned into an invalid, and even die if the seizure is sufficiently serious. But if the neurosurgeon severs the fiber tracts of the corpus callosum, the discharges remain confined to only one hemisphere instead of spreading to the other. In addition, the operation seems to have little effect on everyday mental functions. Neither the patients nor their relatives and friends notice anything different; the patients voice no complaints and act the same as they did before the operation. Despite their seeming normality, the patients are profoundly altered, as can be seen from the results of a couple of experiments.

In one experiment Michael Gazzaniga, who started his career as a student of Roger Sperry, flashed the directive *laugh* to the right hemisphere of a split-brain patient. When the patient laughed, Gazzaniga asked him what he was laughing at. The patient looked over to the experimenters and commented, with another laugh, "Oh, you guys are really something." In this example the left hemisphere (which controls speech) provided an explanation for behavior (laughing). As Gazzaniga explained it, the left hemisphere tried to *interpret* the act of laughing as though it understood the cause of this behavior, which it did not. "It is as if the verbal self [i.e., the left hemisphere] looks out to see what the person is doing, and from that knowledge it interprets reality," notes Gazzaniga.

The second experiment illustrates this tendency for the left hemisphere to "rationalize" our emotions. A patient, V.P., sat before a screen as the researchers directed a movie to her right hemisphere. It was the kind of movie most of us would find upsetting: one man viciously pushing another man from a balcony and then dropping a firebomb on top of him. After watching this grisly film, V.P. couldn't recall anything other than "a white flash, perhaps a few trees but definitely no people." When asked if she felt any emotion about the movie, she replied, "I don't really know why, but I'm kind of scared. I feel jumpy. I think maybe I don't like this room, or maybe it's you, you're getting me nervous." In short, the right hemisphere experienced the appropriate emotions (fear), but in the absence of communication between the hemispheres the left hemisphere could only guess about the cause of that fear. "The verbal system notes the mood and immediately attributes a cause to the feeling," comments Gazzaniga.

A principle emerges here that holds not only for split-brain patients but for the rest of us as well: We are continuously in the process of trying to explain ourselves to ourselves by coming up with plausible causes for our actions. This is especially true in regard to our emotional responses.

For example, when we don't feel good about ourselves, we tend to settle on something or someone as the cause of our bad feelings. After noting that low feeling, our verbal system in our left hemisphere comes up with a cause for it. We may blame it on our spouse, a poor night's sleep, or the promotion that eludes us at work. And if we seek professional help to manage our feelings, we're likely to encounter a similar blaming process espoused by practitioners of the various schools of psychotherapy, who emphasize different and sometimes conflicting "causes" for our feelings.

A mental system sets up a mood that alters the general physiology of the brain, explains Gazzaniga. "The verbal system notes the mood and immediately attributes cause to the feeling. It is a powerful mechanism, and once so clearly seen, it makes you wonder how often we are the victims of spurious emotional cognitive correlations." Which leads to a perfectly sensible question: What is the "true" explanation for our feelings? Are we capable of explaining ourselves to ourselves? I think a certain agnosticism seems justified in response to that question. Any of the "explanations" we wind up accepting as true may merely reflect an arbitrary selection by our left hemisphere among many various possibilities.

"The order we 'discover' is the order of a notational scheme that we project upon the world," as philosopher Roy Sorenson puts it in A *Brief History of the Paradox*.

The Alien Hand

Consider the Wilder Penfield experiments described above. When Penfield electrically stimulated the motor cortex of his operating-room patients, he produced movements that the patient felt no sense of connectedness with. A common description given by many of the patients was that the movements had been somehow "pulled out" of them; they failed utterly to experience themselves as the originator of their own willed actions.

Penfield's observations remained an interesting but inexplicable oddity until 1985, when San Francisco researcher Benjamin Libet measured electrical responses in the brains of volunteers who were requested to make a simple finger movement at any moment of their choice. He found that between 500 and 1,000 milliseconds before the movement a scalp-recorded brain readiness potential (RP) occurred. Not sure what to make of this finding, Libet decided to tinker a bit in the interest of determining the exact point in that 500-msec span when a subject made the conscious decision to move a finger.

Libet asked the volunteers to stare at a clock and at the instant of conscious decision to move the finger also observe and remember the position of a moving black dot on a clock face visible in front of then. Then, after they completed the movement, they reported to Libet the dot's location at the moment they decided to move the finger. Libet correlated that moment with the onset time of the readiness potential and came up with a strange finding: The readiness potential occurred 300 milliseconds prior to the decision to move the finger. Libet wrestled with the profound consequences of his finding.

For instance, Libet could predict the timing of a volunteer's decision to move his finger within milliseconds simply by noting the onset of the readiness potential—even though at that moment the volunteer hadn't yet consciously made up his mind to make the movement. What's more, the volunteer remained convinced throughout the experiment that he had acted at a moment of his own choosing.

Our brain, it appears, knows our decisions before we do. Psychologist Daniel Wegner aptly sums up the take-home message of the Libet experiment: "The experience of consciously willing an action begins after brain events that set the action into motion. The brain creates both the thought and the action, leaving the person to infer that the thought is causing the action."

A common theme can be discerned in the Libet and Faraday experiments: A person may not be the best judge of whether she is actually the originator of her own actions. The table turners were themselves moving the tables, yet were unaware of it. And while Libet's volunteers experienced themselves as acting freely, their actions could be confidently predicted on the basis of an electrical event occurring prior to their conscious decision. Thus it's possible both to be doing something and not know it and to consciously decide about something while remaining unaware that our brain has preempted us.

Neurologists have observed in their patients over at least a century other dramatic abnormalities in the control and awareness of behavior:

• After amputation of an arm or leg, many patients continue to experience the continued presence of the absent limb; some even experience a sensation of the phantom limb moving involuntarily.

• Patients with a condition called alien hand (also known as anarchic hand) experience one of their hands as acting independently of their will. They describe the hand as "foreign" or "uncooperative": "My hand did that" may be offered as an excuse for something the patient is embarrassed about. To take one actual example, the hand may reach toward a nurse and attempt to pinch her. Or the hand may carry out an action directly opposed to the action of the other hand, as in this description by a physiotherapist of her patient, Rocky: "You should have seen Rocky yesterday—one hand was buttoning up his shirt and the other hand was coming along right behind it undoing the buttons."

But the strangest example of behavior carried out in the absence of the exercise of free choice is the phenomenon known as environmental dependency syndrome (EDS), first described in 1986 by the French neurologist François Lhermitte. After damage to the frontal lobe a patient with EDS develops a peculiar and irrepressible urge to imitate other people's actions. In one example we filmed for an episode of the PBS series *The Mind,* two flowers rest in vases on a desk Lhermitte sits behind, across from an EDS patient. When Lhermitte removes one of the flowers from its vase and smells it, the patient imitates the same movement using the other flower. When Lhermitte then starts to comb his hair, the patient takes a comb from her pocketbook and combs her hair. At this point in the film, Lhermitte rises from his chair and kneels in silent prayer. The patient joins him with her head bowed and hands pressed together in supplication. "For the patient afflicted with EDS, the social and physical environments issue the orders to use them, even though the patient himself or herself has neither the idea nor the intention to do so," says Lhermitte.

Lhermitte's patients experienced difficulty with willed control. What's more, they felt compelled to carry out their imitative movements. When asked why they were imitating him, the patients responded that since Lhermitte had done these things they felt that they had to do them as well. This held true even when Lhermitte specifically told them not to imitate his gestures. "On being reminded afterwards that they had been told not to imitate the gestures, their answer was that obviously since the gestures had been made, they must be imitated," says Lhermitte.

The Faraday and Libet experiments along with the EDS patients described by Lhermitte point to a counterintuitive principle of brain operation: Conscious intention and behavior often function quite independently of each other.

In summary, Wilder Penfield's operating-room experiments revealed the existence of a repository of past experiences inaccessible to conscious recall buried deep within the temporal lobes of the brain. Libet showed that the brain decides to act prior to any conscious decision. Roger Sperry refined this insight via his demonstration that each of the hemispheres processes different aspects of the environment. A student of Sperry's, Michel Gazzaniga, took Sperry's research a step further with his discovery that the left hemisphere imposes explanations and meaning on our perceptions and behavior. These four researchers, along with others I have omitted for the sake of brevity, laid the foundations for our understanding of the cognitive unconscious and established a new way of looking at human behavior.

No longer was it sufficient to merely ask, "Why did you do that?" Brain research was leading to the counterintuitive notion that many times we cannot honestly answer that question. Our actions originate outside of our awareness; consciousness plays little part in determining how we respond to many aspects of the world around us. This is a heady and sobering thought: We don't so much make decisions as our brain makes them for us. When we claim conscious ownership of the actions performed by our brain, we act like proud parents of a gifted child, taking credit for the child's brilliant performances even though we only provided the necessary conditions for that brilliance.

This early recognition of the cognitive unconscious led directly to social neuroscience and paved the way for the emergence of the neurosociety. As we shall see in the chapters that follow, this revolutionary way of looking at human behavior is currently exerting a powerful influence in fields not traditionally associated with the brain.

3

The Emotional Brain

Instructed Fear

If you had to select one inspiration for many of the most innovative discoveries in social neuroscience during the last two decades, *conditioning* wouldn't be a bad choice. The concept underlying it is pretty simple. Pavlov started it all by pairing the feeding of his laboratory dogs with the sound of a bell. After the dogs heard the bell a number of times just prior to their feeding, it soon wasn't necessary for them to see or smell food before salivating in anticipation; all they needed was the sound of that bell. Other experiments took on a harder edge, with punishment replacing reward as the motivator. Instead of hearing bells, the animals received electric shocks paired with a sound or a light of a specific color. Soon it wasn't necessary to shock them; the sound or the light alone put them into paroxysms of fear.

According to subsequent conditioning research carried out by Joseph LeDoux at New York University, fear operates along two separate pathways. One path goes directly to the amygdala and activates an immediate response (jumping back onto the curb to

Hey, let's do an experiment on neuroscientists!

avoid being hit by that car that just ran the red light). The second pathway—LeDoux calls it the "high road"—involves the cerebral cortex, the part of our brain where we formulate reasons why we're afraid (those mental images of lying in a hospital bed for several weeks after being hit by that car). Since it takes longer for the second fear pathway to spring into action, we rely on the first pathway in emergencies. And that's a good thing—pause to get the license number of that car coming at you, and you've just won yourself a trip to the hospital or to the morgue. A similar rule holds in the wild: If you dally to ponder the identity of a snake you encounter in the woods, that may be your last exercise in applied taxonomy if the snake is a poisonous one. Better to depend on the direct path involving the amygdala (LeDoux calls it the "low road," or the "quick and dirty path"), jump back, and carry out your classification from a safe distance.

Notice that the conditioning experiments and experiences described above all involve some kind of direct experience. Thanks to our language and imaginative abilities, we humans don't always require direct experience in order to be conditioned. I'm thinking now of a dog kept for a while by a neighbor who lived around the corner from me. Although I had observed the dog from a distance, I had never had any direct contact with it. I had heard rumors of several vicious attacks the animal had made on other dogs as well as at least one aborted attack on a neighbor. The dog's owner had attached a cowbell to the dog's collar so that other dog owners could know of the dog's approach and stay away. Even though I had no direct experience with the dog, I nonetheless anxiously scanned up and down the street whenever I heard the bell.

Although I didn't think about it in such terms at the time, I was conditioned: The sound of the bell activated a fear response. Yet it wasn't the kind of fear response that would have developed had I actually been attacked by the dog. Call it "cold" cognitive processing: no racing heart, no increased breathing, no sweaty palms, simply a low-candlepower dread of running into that dog. Thus in this situation my brain was using the slower-functioning cerebral pathway: "There's that bell, so maybe I'll just stay in the house for a few minutes longer."

Now imagine for a moment that I have a visitor for the weekend and tell her about that dog. If I did that, I would be carrying out what psychologists refer to as the "instructed fear" procedure. As a result of my words, it's likely that my friend will also respond, quite unusually for her, to the sound of a cowbell. In her case, the conditioning is even further removed from direct experience: She hasn't any immediate means of determining the dog's existence other than by what I've told her.

Now let's play a little trick on my visitor. Let's hook her up to a machine and measure some physiological responses and then sound that cowbell. What do you think we'll find? If I have been sufficiently graphic in my descriptions to her about the dangers of the dog, we'll discover a slight but measurably increased arousal, almost as if her brain is now scanning the environment in search for that dog. If we did an fMRI of her, we would also find a corresponding activation of her amygdala. In short, we learn to fear not just what we have encountered but also what others have told us about. Thus the amygdala can be activated in the expression of fears that are only imagined or anticipated but never actually experienced.

As you've no doubt observed by now, many of the new insights about the brain discussed so far in this book run counter to commonsense notions. And fear provides a good example of how our intuitions and the facts on which our intuitions are based can vary greatly. In order to become frightened, it would seem that we would have to be conscious of the cause of our fright. In the example of the vicious dog, both my "cold" cognitive processing and my friend's instructed fear involved conscious thinking about the dog.

But consciousness isn't necessary, according to recent findings on the amygdala.

What It Means to Be Emotional

Imagine yourself looking at a series of pictures of human faces. Your amygdala will remain as inactive as a contented kitten until I show you faces depicting fear. At that point, your amygdala stirs like a lion awakening from its nap. And it doesn't matter whether you consciously perceive the faces. Paul Whalen, a psychologist at the W. M. Keck Laboratory for Biological Imaging, University of Wisconsin, Madison, flashed those fearful face pictures so rapidly that the viewers couldn't possibly be aware of them (the viewers even denied that they had seen anything at all). Nonetheless, the amygdala, in a sense, saw the faces, registered the fear, and sprang into action. In fact, it isn't even necessary for a person to actually see the entire face. The wide-eyed appearance of fear and surprise, resulting from greater visibility of the white portion of the eye surrounding the pupil, is sufficient to activate the amygdala. Cartoonists take advantage of this by drawing wide-eyed expressions when they want to convey fear in their cartoon characters.

And while we're on the subject of cartoons, imagine that you and I are drawing a cartoon sequence in which one character is frightened and another character is angry. How would we do that? Well, we can convey fear in the frightened character by drawing her with a wide-eyed look. But what about anger? For one thing, the most powerful expression of anger always involves direct head-on gaze. That's why most of us are uncomfortable when someone stares at us with great intensity and without a glimmer of a smile. Such conditions activate our amygdala to the extent that it takes a real effort to control our mounting discomfort at the direct gaze of the other person. But if that person looks off to the side, perhaps for just a moment, we feel less threatened and our amygdala powers down.

Neuroscientist Nalini Ambady provides a neuroscientific explanation for the importance of gaze direction. Using the Montreal Set of Facial Displays of Emotion, which depicts actors miming anger and fear, she manipulated the gaze direction of the actors using Adobe Photoshop so that the angry faces looked off to the side while the fearful faces faced the viewer head-on. She found that angry faces seen in direct gaze and fearful faces with the eyes averted are recognized more quickly and accurately than either angry faces shown with averted gaze or fearful faces with the gaze directed head-on. Incidentally, Ambady's findings don't imply that angry people never avert their gaze or that fearful people don't sometimes stare directly at you. We all know from personal experience that anger and fear are sometimes expressed in those ways, but these are exceptions to the rule.

Now let's change our exploration of the amygdala and substitute words for the pictures. Imagine that I'm showing you a series of fifteen words, each displayed on the screen for no more than about 100 milliseconds. At that rate you will only be aware that you've seen some words; you won't be able to identify them. I can improve your recognition by asking you to look for only two words in the list. I can make it even easier for you if I print the two words in red or green ink so they stand out. Under these conditions you will be able to spot those words with one notable exception: If the second word appears soon after the first (only one or two words later), you will miss it. That's because the act of identifying the first word leads to a brief refractory period during which your perceptual processing takes a millisecond's rest, and as a result, you lose the ability to identify a second word. Psychologists refer to this as an "attentional blink."

There is an interesting exception to the attentional blink: It disappears if I include two or more emotionally arousing words in the list. Then it doesn't matter how close the words are together; you will still spot that second emotionally arousing word. "This enhanced ability to identify an arousing second word during the blink period suggests that when attentional resources are limited, emotional stimuli are more likely to reach awareness than neutral stimuli," according to Elizabeth Phelps of the Department of Psychology at New York University.

But—and here's the point—if this test is given to a person with damage to his or her amygdala, there isn't any difference in response to emotionally arousing words and ordinary words. That's because normally the amygdala recognizes emotional significance very early in the brain's processing; but if the amygdala is damaged, that recognition doesn't take place. Before we even know that we're frightened or what we're frightened of, our heart rate increases, our breath comes harder, and our sweat glands begin to work overtime. The emotionally arousing word or picture travels along the "low road" described earlier.

Think of the amygdala as the center of a wheel with the spokes extending to the areas of the brain responsible for sight, hearing, and other sensations. When something emotional activates the amygdala, these areas are activated as well. For instance, the activated amygdala swiftly passes the message to the visual cortex, thus placing it on high alert for anything else in the environment that is emotionally arousing. This two-way communication between the amygdala and the cerebral cortex explains why emotionally arousing words, pictures, or events are exempt from the attentional blink. Such an arrangement increases one's chances of survival: The alerting response leads to increased scanning of the environment in search for additional potential perils.

Fortunately, those of us with a normal amygdala (actually amygdalae, the proper Latin term for the pair of structures, one on each side of the brain) have emotional responses in sync with our thinking. It's neither useful nor necessary to split hairs about whether we are acting emotionally or according to the dictates of pure reason. Indeed, distinctions between emotion and reason presently enjoy an undeserved prominence in our concepts about ourselves. Since this is an important insight revealed by social neuroscience, let's take a moment to explore it further.

Take memory, for instance. Two brain structures are crucial for the formation of memories. The hippocampus, a seahorse-shaped structure located near the temporal lobe, is responsible for the initial encoding of all new information. If I've just given you my telephone number, the hippocampus takes down that information and holds it in temporary storage. Damage to the hippocampus, as happens in Alzheimer's disease, results in failures of encoding, hence the memory problems associated with this illness.

The second structure important in memory is the amygdala. It comes into play when the memory involves a strong emotional component. You remember quite vividly where you were on September 11, 2001, because the horrific events of that day aroused your emotions and activated your amygdala. (Incidentally, this emotional activation doesn't always lead to an accurate memory of what happened, as we shall see in Chapter 8.)

Since the hippocampus is taken up with the learning of explicitly or consciously remembered events, the term *episodic* is used to describe hippocampal-dependent memory.

Most episodic memories are pretty bloodless and unemotional—they involve the who, what, when, and where of things. As a result, we don't always remember the exact circumstances surrounding our learning a specific fact. Do you remember the occasion when you learned that Abraham Lincoln was our nation's sixteenth president? Most unlikely. But if your school burned down the afternoon of the day you learned that fact about Abraham Lincoln, you might well remember the circumstances. That's because that item of historical information will have an emotional component associated with it that other memorized historical facts lack. The school fire was sufficiently emotionally arousing (you were upset, sad . . . perhaps glad?) that, thanks to your amygdala, many of the things that happened that day, including learning about Abraham Lincoln's position in the presidential chronology, are tinged with emotion. If you later experience amygdala damage, that emotion will disappear: The sixteenth president is now only one of many historical facts.

Fortunately, for most of us, the hippocampus and the amygdala perform wonderful duets together. Since most of the things we learn are pretty prosaic, they're taken care of by the hippocampus, with the amygdala not playing much of a role (as in, say, remembering a restaurant telephone number). But when danger, threat, embarrassment, and other messy situations come into play, the amygdala kicks in with a vengeance.

So when it comes to emotion and reasoning, what do we conclude about the hippocampus and amygdala? Are they in the service of reason or emotion? The correct answer is both. Think back to my weekend guest, whom I told about a vicious cowbell-wearing dog. Now suppose I told you that she suffers from amygdala damage. What do you think will happen now when she hears the cowbell?

Actually, her behavior will be just the same as before, and she'll steer clear of the dog. Yet her physiological testing will show quite different responses. Most striking is that she will fail to show the normal indicators of fear. In other words, in spite of her awareness of the dangerousness of the dog, nothing about her bodily responses gives any clue that she's concerned about the dog. In a way she's acting rationally (staying away from the dog), but her rationality isn't based on factors such as the fright felt by me and the other neighbors.

Neuroscience is suggesting here that we must change our ideas about reason, rationality, and what it means to be emotional. That isn't going to be easy because of long-held assumptions within our society. For instance, if you serve on a jury, you will be asked to base your decision of guilt or innocence upon how a "rational person" would act under the circumstances leading up to the crime. In other words, you are requested to put emotions aside and apply a standard based on pure rationality. But social neuroscience shows that such a decision doesn't make sense: Thinking and emotionality are inextricably intertwined.

4

How Our Brain Constructs
Our Mental World

Von Hemholtz's Darkroom Experiment

L ook up from this book and focus your attention for a few
seconds on the first thing that meets your eye. Then return
to the book. When you looked up, you didn't have any problem
understanding what I meant by *attention,* did you? One moment
you were engaged in reading a sentence, and the next moment
you had shifted your attention and you were looking at . . . whatever. It was as if you had directed the beam of a flashlight from one
portion of a darkened room to another. Yet attentional shifts don't
necessarily involve eye movements. You could just as easily have
shifted your attention by simply thinking of something you had to
do later today. But what is it exactly that you would be shifting at
such a time? This question tantalized the eminent nineteenth-
century German scientist Hermann von Helmholtz, who carried
out an experiment on attention.

Helmholtz was intrigued with the question of how much information a person could process in a brief period of time. To find
out, he used a flashbulb to briefly illuminate an otherwise dark

scene consisting of letters painted onto a sheet suspended at one end of his lab. When Helmholtz triggered the flash of light in the dark it provided a short-lived illumination of the letters. He immediately discovered that he couldn't take in all of the letters during the brief illumination; he could see only some of them. But if he decided ahead of time which portion of the screen he would attend to during the illumination, he could easily discern those letters despite his continued inability to perceive letters elsewhere on the screen.

Helmholtz was paying what psychologists now refer to as *covert visual attention* to a chosen region of the sheet of letters: "By a voluntary kind of intention, even without eye movements, . . . one can concentrate attention on the sensation from a particular part and at the same time exclude attention from all other parts."

Today neuroscientists can illustrate what's going on in the brain during Helmholtz's ingenious experiment. When you focus your visual attention on something in your immediate surroundings, the blood flow immediately increases in the visual areas of your brain. But this increase in activity isn't just a generalized increase; it occurs in a highly specific pattern. If you attend to something off to your left, your brain's right visual area is activated; if you attend to something off to your right, it's the left visual area that comes alive. (Remember that the visual hemispheres on the two sides of the brain scan the opposite visual fields.)

Think for a moment what such findings imply. As you direct your visual attention to something in your surroundings, your brain has already begun—thanks to your intention to look at one thing rather than another—to selectively focus on that one aspect of the world in front of you. Thus, in the words of George R. Mangun of the Center for Mind and Brain, University of California, Davis, "Changes in visual brain processing significantly affect how we perceive and respond to the world around us."

Magicians have known about this intention-attention link for centuries and take advantage of it via the technique of misdirection. If a trick "works," it's often because the magician has successfully fooled his audience into purposely (intentionally) focusing their attention on an unimportant aspect of the trick, thus preventing them from seeing what he's actually doing.

"Thinking Is for Doing" — Wm. James, p. 57

Not only intention and attention but all other states of mind and body are related to and oftentimes determined by our brain. As an example of my point, say the word *zeal* out loud. Now without changing the position of your lips and mouth in any way, simply imagine saying it. Hold that imagined feeling. Now open your mouth as wide as you can and imagine once again saying *zeal*. Feels quite different, doesn't it?

This simple exercise underscores an important point: Our mental processes are sufficiently tethered to our bodily senses that we have difficulty with situations when the brain and other parts of the body aren't in sync. You'll experience difficulty in any mental activity when your body executes movements that thwart it: It's hard to think critically while slumped in an armchair, hard to meditate on compassion while punching a bag.

Here's another example. Clench your fists, hard, and grit your teeth. At the same time imagine yourself pushing a heavy piece of furniture across the living room. Then quickly change the thought to lying on a beach in the Caribbean listening to gentle waves. Notice how much harder it is to maintain that thought with those clenched fists and that tightened jaw? Now unclench your fists and loosen your jaw and return to the image of pushing that furniture. That doesn't feel quite right either: We don't push furniture with hands and jaw relaxed.

Next, sit for a few moments with your forearm flexed as if

you're about to pull something toward you. Now imagine me showing you a series of items or speaking a series of words and then asking you how you feel about them. I then ask you to repeat the exercise with your arm held straight out in a fully extended position. Would it surprise you—as it did me—to learn that you'll tend to like the various items you encountered while holding your arm in the flexed position, and dislike the items heard while your arm is extended? That's what was found in a test measuring the association of arm posture and attitude.

Flexing the arm, of course, is a motion we all carry out when we're pulling something toward us, "embracing" it; straightening the arm, or "strong-arming," in contrast, is what we do when we want to push something away. For those requiring more convincing evidence of the connection between our personal evaluations and our bodily positions, the experimenters repeated the experiment, but this time with a twist: Half of the participants in the experiment were asked to push a lever away from them if they reacted positively to a particular word but pull it toward them if the word gave rise to negative associations, while the other half of the participants were told to do the opposite, pulling forward with positive words and pushing away with negative words. Overall, people were faster to respond to positive words when they were pulling instead of pushing the lever, and faster to respond to negative words when they were pushing rather than pulling the lever.

The experimenters tried the experiment again, only this time they didn't say anything about likes or dislikes. Half the volunteers simply responded by pushing the lever as quickly as possible whenever a word appeared; the other half of the participants in this reaction time experiment pulled the lever at the instant they became aware of the word. Again, those pushing the lever reacted more quickly to negative words, while the lever pullers responded faster to positive words.

"Immediately and unintentionally a perceived object or event is classified as either good or bad, and this results, in a matter of milliseconds, in a behavioral predisposition toward that object or event," according to Yale University psychologist John Bargh, who carried out the research just described.

Think for a moment about the usefulness of such an arrangement. Thanks to these automatic responses, it's not necessary to consciously evaluate everything that's happening from moment to moment. Rather, our bodily movements automatically bring us closer to positive events and experiences but increase the distance between negative ones and us. This is especially helpful when our conscious mind is otherwise engaged in thinking about other matters, such as our presentation at tomorrow's weekly staff meeting.

This intimate association between body and thought intrigued the nineteenth-century psychologist William James, who emphasized the intimate association that exists between thinking and action. In 1890 James wrote, "It is a general principle of psychology that consciousness deserts all processes where it can no longer be of use. . . . We grow unconscious of every feeling which is useless as a sign to lead us to our ends." Unlike his brother, the novelist Henry James, William James favored tightly compressed aphorisms over lengthy and baroque paragraphs: "Thinking is for doing," he wrote. These four words simplify without oversimplifying the notion that merely thinking about doing something increases the likelihood that one will actually do it. While this seems a fairly commonsense notion—most of us can readily bring personal examples to mind—James developed it a good bit further. Indeed, he took the notion quite literally and argued for the then-maverick view that thinking about doing something activates the same brain regions that come into play when one actually does it.

A hundred years later PET scan studies confirm James's proposal. Starting in the early 1990s, neuroscientists provided PET scan evidence that thinking about a word or about carrying out a movement activates the same area in the anterior cingulate that is activated when actually saying the word or carrying out the movement.

"These studies support the notion that thinking about something and doing it are neurologically similar. And the two activities activate the same regions of the brain, suggesting they share representational systems," says Duke University psychologist Tanya L. Chartrand, an expert on the link between thinking and doing.

Chartrand's comment is important because it's in line with the advice of psychotherapists and motivational experts who suggest imaginational exercises as the first step toward self-improvement. They advise envisioning yourself as the person you want to become, or changing a situation that's troubling you by imagining that change as a prelude to making it happen.   Hmm...

Mirror Neurons

Although we like to think of ourselves as independent and self-actualizing, our thoughts and behavior are powerfully influenced by other people's actions. This holds true even at the level of simple observation. When we watch another person move, our observation of the movement activates those areas in our brain that we would use if we moved the same way. This was first discovered in macaque monkeys, where "mirror neurons" in the prefrontal cortex respond both when the monkey grasps a peanut and when it watches another monkey grasp it. Even hearing sounds suggestive of a monkey grasping and then breaking a peanut activates the mirror neurons. This suggests that the mirror neurons for vision and hearing aren't just coding movements and sounds but rather goals and meanings: What is the monkey doing? Further, mirror neurons can be trained. If a monkey is taught to rip paper or perform some other action that doesn't come naturally to the animal, specific mirror neurons will start to fire at the mere sound of ripping paper.

A similar perception-action matching system exists in the human brain. Imagine yourself watching me reach out and grasp the cup of tea that now sits on the small table next to my word processor. As you observe my hand reaching for the cup, the motor cortex in your brain will also become slightly active in the same areas you would use if you reached out to pick up that teacup. Further, if you watch my lips as I savor the tea, the area of your brain corresponding to lip movements will activate as well.

No, that doesn't mean you can taste my tea. But it does mean that I'm directly affecting your brain as you watch me go through the motions of drinking my tea. In such a situation the neat division between you and me breaks down and we form a unit in which each of us is influencing the other's actions at the most basic level imaginable: I am altering your brain as a result of your observations of me, and vice versa.

Notice that it isn't necessary for you to consciously think about the movement in order to get your brain working. Merely observing me move my hand toward the teacup will activate those portions of your brain that would come into play if you actually moved your hand. But if I move my hand toward the teacup for a purpose other than sipping tea, the mirror neurons fail to fire. We know this because of a clever experiment carried out by a team led by Marco Iacoboni of the University of California, Los Angeles.

In this test of mirror neuron responses, volunteers watched video clips taken before and after a tea party. The "before" clip showed a steaming teapot and cup placed alongside a neatly arranged plate of cookies. The "after" clip depicted an empty teapot, scattered cookie crumbs, and a used napkin. At the conclusion of both video clips a hand reached in from off-camera and grasped the teacup. Since the hand motion was identical in each clip, only the context suggested two different intentions: drinking the tea in the "before" clip versus tidying up in the "after" clip.

As fMRI scans of the volunteers' brains showed, brain activity in these two situations differed markedly. The greatest activity in the right frontal cortex (known from previous research to process mirrorlike responses to hand movements) occurred while the volunteer watched the grasping movement associated with drinking the tea. Thus mirror neurons are affected not just by motion but also by the motivation behind it, according to Iacoboni.

Think for a moment of the implications of this. You can activate my brain if you can attract my attention enough to get me to watch what you're doing, and vice versa. Thanks to the mirror neurons in each of our brains, a functional link exists between my brain and yours.

Nor does any actual movement have to take place in order for this mutual influence of one brain on another to take place. For instance, imagine someone reading to you the following list of words: *plain, rip, geography, stomp, Ireland, wistful, lift.* If you are listening to the words while in a PET scan, the action verbs *rip, stomp,* and *lift* will activate areas of the brain normally engaged if you were actually ripping, stomping, or lifting. Called into play would be the dorsolateral prefrontal cortex, the anterior cingulate, and the premotor and parietal cortices. The same thing would happen if you were watching someone else rip, stomp, and lift, if you spoke the words aloud, or if you dredged up those words from memory.

Valeria Gazzola of the BCN Neuro-Imaging Center in Groningen, the Netherlands, has spun the mirror neuron concept in a more intimate direction. In her experiments on auditory empathy she discovered that when a person listens to a sound associated with an action such as kissing, the act of listening activates the same area in the premotor cortex that would come online if that person actually kissed someone: The kissing sounds activated areas of the premotor cortex controlling the mouth movements associated with kissing. Certainly Gazzola's findings are in sync with our everyday experience. As Gazzola puts it: "If in a hotel room, we hear the neighbor's bed squeaking rhythmically, we can't help hearing more than just a sound."

Part of that intuition comes from the capacity of mirror neurons to distinguish between biological and nonbiological actions. Beds don't move on their own and, with one notable exception, usually don't squeak rhythmically. Monkeys make a similar distinction when another monkey is involved in an action compared to a machine carrying out that action. It will grasp something when it sees another monkey do it, but remains unmoved when pliers or a mechanical tool performs the same action. Human infants show a pattern much like this. At eighteen months an infant will imitate and even complete an action made by a human but will fail to imitate a robot making the same movement. This also holds true for sounds. Infants older than nine months can learn new speech sounds to which they have never been exposed, but only if the new sounds come from a real person. Learning the new sound doesn't occur at this age if the infant hears the same word on a tape recorder or video.

In adults, speaking action words activates the same brain structures as actually carrying out the actions (the verb-behavior link, as neuropsychologists refer to it). That linkage is one of the reasons we become annoyed when someone asks us a question while we're trying to concentrate. The question automatically activates those parts of our brain involved in formulating the answer to that question. We're annoyed because of the conscious effort we must exert not to become distracted—that is, direct our attention away from our current thoughts in order to formulate a response to the question. In such situations we learn firsthand that although our brain can carry out many processes simultaneously, the focus of our conscious attention is limited to a few things at a time.

Mental imagery is actually an offshoot of our capacity for activating those mirror neurons. Think for a moment about a pleasant experience you had, say, during your last vacation. I'm thinking at the moment of the view of the beach from the patio of my hotel room at Maui in Hawaii. Although I can "see" the scene very clearly in my imagination, the experience obviously isn't the same as actually being there. From the point of view of physical location, thousands of miles separate the two experiences. But what about the representation of these two experiences in my brain? How does my creation of a mental image within my brain differ from the original experience? Actually, not nearly as much as you might expect.

Over the last decade neuroscientists have discovered that the visual imagery employed in an imaginative re-creation of an earlier experience shares important features with the original visual perception. As I think about the beach many (not all) of the same areas will be activated in my brain that were active when I was physically there. And the time that it took me to scan the view of the beach from my patio is equal to the time it takes me now to mentally imagine the same scene. Moreover, if you ask me a question about the beach, my brain will direct my eyes to look downward, just as it would if I were once again looking down from the patio. A question about how the beach appeared from the right of the patio will provoke an unconscious shift of my eyes to the right (or to the left if you asked about how the beach appeared from that direction).

Thanks to mirror neurons, we can mentally rehearse physical activities without actually doing them. If you've learned a physical exercise and then later mentally rehearse it, you will induce an increase in muscle strength comparable to what would happen if you actually did the exercise. What's more, your heart rate and

to what degree?

breathing frequency will increase linearly with the increase in the imagined effort: the greater the exertion you imagine yourself making, the more your heart and breathing rates will increase. If you are put into an fMRI scan, many of the same areas will activate as when you actually did the exercise (the primary motor cortex, the premotor cortex, the SMA, the basal ganglia, and the cerebellum).

thank you!

Not everyone shows the same degree of mirroring. When observing someone playing the piano, skilled pianists show stronger motor activation than the musically naive. But if they watch random finger movements instead, the brain responses of the pianists are indistinguishable from people with no special musical expertise or interest. A similar situation exists among dancers. In an ingenious study Bentriz Calvo-Merino of the Institute of Movement Neuroscience, University College London, compared the brain activation patterns of ballet dancers and capoeira dancers (capoeira is an Afro-Brazilian martial dance first developed by slaves in Brazil more than four hundred years ago as a means of fighting back against the slave owners). She found greater brain activation when the dancers viewed movements they had been trained to perform compared to movements they had not. That is, the ballet dancers showed stronger brain activation while observing ballet movements than when observing capoeira movements, and among capoeira dancers, the greater activation occurred while watching capoeira.

In response to this demonstration of the power of imagery for elaborate movement sequences, coaches and trainers have incorporated various imaging exercises into athletic and physical fitness training programs. Obviously, this reliance on imaging can only go so far. That's why the world is still waiting (and will continue to wait, I expect) for the first winner at Wimbledon whose training has consisted only of mental exercises. While thinking and doing activate the same brain areas, the winning of a tennis game, as any regular player knows quite well, requires physical and not just mental effort. *Because physical involves many more (and more finely differentiated, detailed) movements and micromovements*

So think about those mirror neurons the next time someone talks to you about the power of human imagination. Imagination involves the activation of sensory and action centers in the brain. Not only can we now locate it within specific areas of the brain, but we can also quantify it. And for the most part *(not entirely)* imagination is a power that we control: We consciously invoke the images in our imagination or, on those occasions when an image spontaneously springs to mind, we can consciously suppress it. Further, we can "build up" our imaginative powers just as we do our muscular strength.

With this as background, let's further explore how our most elementary perceptions are grounded in social interactions.

The Romance of the Circle and the Little Triangle

Look at the two dot patterns on page 65. Any suggestions as to what they might represent? Difficult to say, isn't it? They could represent anything or nothing. Yet if I were to set those dots into motion, you would readily recognize them as representing the human body (see the figures on page 65). We have the ingenious psychologist G. Johansson to thank for that insight.

More than thirty years ago Johansson filmed actors dressed in black with white dots attached to their joints as they moved against a black background. He discovered that by watching the dots people could correctly identify the walking or running motions of another human or an animal. In the case of human movement, they could also correctly infer on the basis of the pattern of the moving dots the subject's gender and specific identity (if the person with the attached dots was someone previously known to them).

An example of a biological
motion stimulus. (Adapted from
G. Johansson, "Visual Perception
of Biological Motion and a Model
of Its Analysis," *Perception and
Psychophysics* 14 [1973], with
permission from the publisher.

Such a highly developed ability no doubt stems from the fact
that human movement is the only kind of motion that we humans
both produce and perceive. And most of us are really good at analyzing it. We can identify a friend two blocks away based on our
recognition of his gait. But if that gait is modified in any way (perhaps by something as simple as the favoring of one leg as the result
of an ankle strain) we may be uncertain of the person's identity
until we move closer. A similar fall-off in performance occurs with
the Johansson point light display—change the configuration of
the moving pattern of dots and the percentage of correct identifi-
cations drops significantly. In all cases our accuracy is greatest
when we are looking at our own movements: When observing
the patterns of moving dots, we identify ourselves first, and then
others.

From an evolutionary point of view, our brain's specialization
for movement makes a good deal of sense. Survival in the wild

demands the ability to identify potential predators on the basis of their movement patterns. Better to react to threatening movements than to wait until we can distinctly envision a predator. Moving-dot tests in animals conforms this movement-over-form bias. Cats can distinguish the movements of other cats in the dark when watching a Johansson-type biological motion display. In short, both in the wild and in the laboratory, motion detection trumps form detection. But what part of the brain is responsible for this bias?

Brain imaging studies have pinpointed the superior temporal sulcus (STS) as the region that is most active when someone looks at moving-dot animations. The STS, shown in the diagram on page 16, is one of the three or four most important brain areas in social neuroscience. It serves as the convergence point for two separate visual streams starting at the occipital (visual) cortex. The top stream, informally known as the "where" pathway, is concerned with motion; the bottom stream, the "what" pathway, enables us to recognize what we're seeing at any given moment. And while both streams converge on the STS, they don't do so simultaneously: Motion information arrives first (by some 20 milliseconds).

We personally experience this precedence of movement over form on a daily basis. If you're sitting in a restaurant talking with a friend, you're likely to find your gaze repeatedly darting in the direction of that basketball or football game playing on the television over the bar. This will happen even though you may have no interest in sports. It's the motion of the players that grabs your visual attention. Advertisers routinely employ motion-filled ads in order to more easily grab the attention of viewers. Special effects in the movies exert a similar capture of your attention, one of the reasons serious drama involving subtle interpersonal interactions doesn't usually feature a lot of special effects lest attention be diverted from the story.

Not only do we preferentially respond to movement and recog-

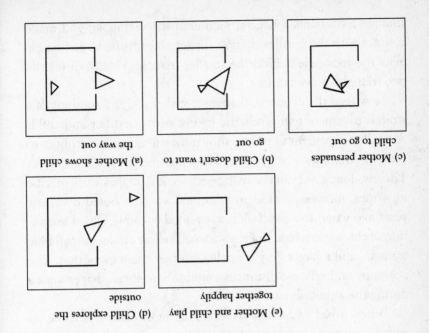

(a) Mother shows child the way out

(b) Child doesn't want to go out

(c) Mother persuades child to go out

(d) Child explores the outside

(e) Mother and child play together happily

nize ourselves and other people from movement patterns alone, but we also tend to project human characteristics onto impersonal objects. For instance, look at the five diagrams above. They are stills taken from an animated forty-second film entitled *Theory of Mind*. As you can see, the stills depict two triangles moving about within an open rectangle. The triangles are at various times either inside or outside the open rectangle. Now, without looking at the captions (which I have printed upside down in order to make that easier for you), how would you describe what is going on as the animation progresses from (a) to (e)? (Hint: Does there seem to be any relationship between the movements of the two triangles? Or is their movement random and arbitrary?)

If you answered that it was random and arbitrary, look again. In (a) the small triangle is placed deep within the rectangle at the moment the large triangle exits the rectangle. In (b) the large triangle has changed course, reentered the rectangle, and made contact with the smaller triangle. In (c) and (d) the large triangle exits the rectangle behind the smaller triangle. Finally, in (e) the two triangles make contact.

Now read the captions. Perhaps it strikes you as something of a stretch to equate two geometric figures with a mother and child? Actually, 93 percent of people shown such a sequence attribute to the two triangles precise, purposeful interactions of some type. This tendency to imbue moving geometrical shapes with specific identities, motives, and even emotions was first noted over fifty years ago when two psychologists created a simple filmed animation of three geometric shapes—a small yellow circle, a small blue triangle, and a larger gray triangle—and set them in motion. The positions and relative distances among these three shapes varied during the animation.

When asked what they were seeing, most observers came up with imaginative narratives such as a romance between the circle and the little triangle, the big gray triangle trying to abduct the circle, the small blue triangle putting up a fight, or the circle and the small triangle escaping into the house (the rectangle) and embracing.

Other descriptions are possible for the two triangles and the rectangle we looked at a moment ago from the *Theory of Mind* film. A husband is trying to interest his wife in going outside on a beautiful spring day. When she demurs, he reenters the house, talks further with her, and succeeds in persuading her to come outside with him, where they embrace. More sinister explanations are also possible. Stills (c) and (d) can be interpreted as a forceful expulsion of the wife by the husband, followed by restraint applied outside the house (e). I'm sure you can readily come up with yet other interpretations. The point isn't the "correctness" of the interpretation; as the experimenters routinely tell their subjects, there is no "right" or "wrong" interpretation. The real issue involves what happens in the brain when these human attributions are applied to moving geometrical forms.

Four main regions are activated: the medial frontal cortex, the superior temporal sulcus (the junction between the temporal and parietal lobes), the temporal regions at the base of the brain near the amygdala, and the occipital gyrus. These are the same areas activated when we monitor our own actions ("Am I behaving rudely here?"), perceive the actions of others (identifying an acquaintance from a distance on the basis of his gait), and attribute mental states to other people on the basis of what they say and do ("Judging from his tone of voice, Frank sounds depressed; and he's sitting slumped in his chair as if he's carrying the weight of the world on his shoulders.").

Here's the summary of the meaning of all this, according to Christopher Frith of the Wellcome Department of Imaging Neuroscience at University College London: "The ability to make inferences about other people's mental states evolved from the ability to make inferences about other creatures' actions and movements." In other words, we evaluate other people's mental states using the same circuitry we employ when observing their movements.

I'm thinking about the Johansson effect while sitting at an outdoor table of a local restaurant, where I'm drinking some iced tea and watching the parade of characters that walk by. The fun associated with people watching involves the same kinds of challenges posed by the moving-light display: trying to deduce other people's intentions and moods based on analyzing dynamic visual information about their movements. One person is ambling along, seemingly taking all the time in the world to get to a vaguely intended destination. Another flits through the crowd with a worried look on his face that signals a firm determination to get somewhere fast. Such "reading" of other people based on their bodily movements is one of the most important perceptual activities we engage in. Our social survival demands that we become adept at mentalizing: recognizing in other people mental states such as anxiety, sadness, impatience, disapproval, and so on.

Not everyone is equally good at reading people's mental states from their bodily postures and movements. Nor are people equally skilled in expressing their own emotions via subtle gestures, facial expressions, or tones of voice. Let's examine what happens within the brain during mentalizing.

Sally, Ann, and the Bank Robber

As an illustration of mentalizing, look at the cartoon sequence on page 71, known as the Sally-Ann test, which depicts the behavior of two little girls, Sally and Ann. In the first frame, Sally places a doll in her carriage. When she leaves the room, Ann takes the doll and places it in her box. In a moment Sally will return to the room. Now please answer the following three questions: Where is the doll? Where did Sally put the doll? Where will Sally look for the doll?

If you're old enough to be reading this book, you'll have no difficulty answering those questions. Starting at age four, a child will correctly answer that Sally will look for the doll in her carriage because that is where she mistakenly believes it is; a child of three, however, will answer that Sally will look in the box because that's where the doll actually is. But the test isn't about the location of the doll, but about what Sally is likely to believe in regard to its location. Three-year-olds fail the test because in order to come up with the correct answer they have to mentally put themselves in Sally's place, and they can't yet do that.

Neuroscientists use the Sally-Ann test and other false-belief (FB) tests to evaluate one person's ability to attribute mental states to another person. This is a gradually developing process that occurs unevenly within the population. By age five the vast

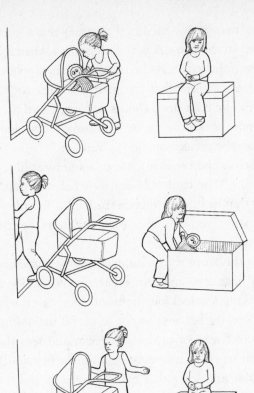

majority of us are able to pass much more challenging FB tests by entering into another person's mind by proxy and making an accurate approximation of how that person is thinking or feeling under particular circumstance (for example, when Sam slips on the dance floor, we wince at the embarrassment that we imagine he must be feeling). But while such ability to put oneself into the shoes of somebody else may well be unique to our species, it isn't universal, even among adults. Autistic people have great difficulty doing this—80 percent of them fail the Sally-Ann test.

Here are two stories cited by Frith that test a person's ability to mentalize. In the first, "A burglar has just robbed a bank and is escaping down the street. As he turns a corner a policeman on his beat who doesn't know that the man is a bank robber sees him drop a glove. The policeman shouts out, 'Hey, you! Stop!' in order to tell the man about the glove. The bank robber looks over his shoulder, sees the policeman, and surrenders."

If asked to explain that story, you would most likely say something like this: "The bank robber falsely believes that the policeman knows that he has just robbed the bank." In order to come up with that answer, you mentalize the story from the bank robber's point of view.

Contrast this story with a similar one where mentalizing plays no role: "A burglar is about to break into a jewelry shop. After skillfully picking the lock on the front door, he crawls under the electronic detector beam so as not to set off the alarm. Moments later he opens the door of the storeroom and sees the glittering gems on their trays. As he reaches out for the first tray he steps on something soft and furry that emits a screech and then scurries toward the shop door. At this instant the alarm sounds."

In this second story the mental state of the thief plays no part in the correct interpretation of what happened: When stepped on, an animal lets out a cry of pain, speeds toward the door, and in the process sets off the alarm. Physical and not mental causality determined the action of the story.

Now imagine yourself listening to those two stories while lying inside an fMRI. During the jewelry shop story, where mentalizing played no role, activity would appear in the superior temporal sulcus (STS) and the cingulate cortex. The story of the bank robber and the policeman would activate those same areas plus the medial prefrontal cortex (MPFC). Indeed, the MPFC springs into action whenever we direct our attention inward and think about ourselves, or outward and think about others.

"You're Fine, Thanks. How Am I?"

On a typical day we encounter dozens of people who ask us, "How are you?" Although social mores require that we respond, "Fine, thanks" (to everyone except our doctors), that question may on occasion provoke in us a more candid reflection: "Well, come to think of it, how *do* I feel?"

"Our social world prompts us to reflect not just upon our own feelings, but the feelings of those around us as well," says Kevin Ochsner, director of the Social Cognitive Neuroscience Laboratory at Columbia University in New York. "As we interact with colleagues, compete with opponents, or watch our friends and family experience their own ups and downs, having insight into the feelings of others enables us to understand what they value, how they feel about us, to offer appropriate support or gain competitive advantage, and to predict their future behavior."

Self-examination of our feelings from moment to moment can help us to identify important aspects of ourselves and other people. It can also provide a means of identifying situations worth seeking or avoiding.

Think back to an experience in your life when you were forced into contact with someone who seemed unable to respond to your feelings. As clearly as you explained yourself, and despite all of the bodily signals that accompanied your explanations, the person just didn't get it. Remember how miserable and misunderstood you felt? If the situation worsened even more, you may have concluded that you were a target for harassment and responded with annoyance. This tendency to respond negatively to people who fail to "read" us begins early in life.

According to infant researcher Elizabeth Meins of Durham University in England, our sensitivity to other people's ability to correctly read our emotions starts in infancy. In a compelling study

she videotaped more than two hundred mothers as they interacted with their infants at ages eight, fourteen, and twenty-four
months. Infants with mothers who were adept at reading their
moods progressed the fastest by age two in language and play
skills. These "mind-minded mums," as Meins refers to them, can
intuitively and correctly read their children's emotions and states
of mind, accurately judging when their infant was feeling contented, upset, or tired. No such gains occurred among infants of
mothers who misread their babies, claiming, for example, that
their child was upset or tired when the child showed no signs of
this. The mother's education or social status didn't account for
mind-mindedness—half of the women in the study had ended
their formal education at age sixteen and were either unemployed
or working in unskilled or semiskilled jobs.

Admittedly, Meins's study is a small one, yet it's consistent
with our everyday experience: We thrive when we're dealing with
people who understand us, people who appreciate and respond
correctly to how we're feeling. And while some people are good at
reading us, others seem perfectly oblivious to our feelings and
mental states.

Based on what we've said so far about the medial prefrontal
cortex, you won't be surprised to learn that social neuroscientists
have recently identified that structure as the center for "mind-
mindedness."

The MPFC is responsible both for our concept of how we feel
from moment to moment and for our ability to intuit the feelings
of other people. Indeed, the MPFC springs into action even when
we only imagine experiencing an event. The same thing occurs
whenever we imagine ourselves in the shoes of another person.

Imagine yourself taking part in an fMRI experiment involving
a rapid shift in your attention from a word game to a briefly presented target. After a few minutes you are told that you are being
observed via a video camera during some parts of the experiment

but not others; further, you will be informed when the camera is on and when it is off. What effect do you think the camera-on segments will have on your experience, in contrast to the camera-off periods?

If you're like the subjects who participated in this study, you will become more self-conscious when the camera is on. "I feel like people are watching me" and "I wonder how I look" were typical comments. Your reaction time will also increase when the camera is on. Presumably, your interior dialogue about your performance, accompanied by your evaluations and self-criticism, take a toll on the rapidity of your response. But the really interesting findings appear in your fMRI.

During the camera-on periods your MPFC will activate, followed by deactivation during the camera-off segments of the experiment. At one moment you're concentrating on doing well in the experiment (camera off), and the next moment (camera on) you're concentrating on yourself—specifically, how you will appear to others who may be watching you on a monitor. The MPFC springs into action as you shift your focus from the word game to how other people may be judging your performance in the game.

"The region of the MPFC associated with mentalizing is activated whenever we are attending to our own mental states or the mental states of another person," says Christopher Frith. We wince if we are seeing someone in pain and become anxious when looking at a picture depicting a scene of horror, but we are responding to our own mental representations, not to the pain or the scene. "Activity in the MPFC is connected with the creation of these representations of the world. Thus mentalizing is about representing our own thoughts, feelings and beliefs as well as representations of the mental states of other people. And what determines our behavior is not the state of the world, but our beliefs about the state of the world," notes Frith.

Think for a moment about the implications of Frith's statements. Our thoughts and feelings about ourselves and other people are processed in the same areas. This confirms what sages and religious leaders have been saying throughout the ages: We shouldn't think of ourselves only as isolated components in an impersonal social network. Rather, like it or not, we are deeply social creatures. Thanks to the MPFC, we can mentally transcend our limited personal perspective and see things from another person's point of view. It's the MPFC that makes it possible for us to attribute mental traits to other people, sometimes even to objects that we know can't experience mental traits (those moving triangles).

One more tidbit about the MPFC: It is among those brain areas having the highest baseline activity at rest. Even when we're thinking of nothing in particular, our MPFC is perking along in what Marcus Raichle, one of the pioneers in human brain imaging, refers to as a "default state." And since we're social critters, that makes a lot of sense. Every time you imagine how another person may feel in a given situation, your MPFC is called into play. As an example, consider this short description of a person's unintentional violation of ordinarily accepted social behavior: "Joanna is invited for a Japanese dinner at a friend's house. While tasting the first course she chokes and spits out the food." After reading this description, volunteers in an experiment were requested to try to imagine how the other people at the dinner table felt while witnessing this turn of events. Their fMRIs were then compared with fMRIs of people who read a description of the same dinner party that omitted the description of Joanna's behavior. Only the Joanna details elicited activity in the MPFC (along with the temporal lobes and STS). That fMRI pattern suggests that simply reading a description of the embarrassing event led to mentalizing about the responses of the other members of the dinner party.

The Illusion-of-Truth Effect

As the above examples make clear, context is important. Within the brain, things are always evaluated within a specific context. So, as marketers and public relations specialists have long recognized, it's important to always retain control of the context in which information is first learned.

Controlling context is important in order to take advantage of the fact that most of us often forget the context in which we first heard a piece of information. Did we hear that information from a friend, or did we read it in a newspaper, or did we hear it on television? A similar situation arises when we hesitate for a moment before repeating a joke currently making the rounds: "Have I told him this before? Indeed, am I absolutely certain I didn't first hear it from this person?" Such uncertainties make us highly suggestible.

Here's how this works. In an attempt to compensate for this loss of context, the brain unconsciously assumes that familiar information is true information: "I've heard that before, so there must be something to it." As a result of this "illusion-of-truth effect" (the technical term employed by psychologists), repetition of negative information is very likely to bring about the opposite effect to what is intended. The brain doesn't do well when it comes to remembering negative information. $= 1986$

I first became aware of this principle twenty years ago from Ann Buchwald, then my literary agent. Ann and I were watching a televised press conference called by a Hollywood star to deny "persistent rumors" about his sexual orientation. "Never call a press conference to deny something," Ann told me. "A few months from now the only thing people will remember about the press conference was that he stood up there and talked about homosexuals. In their minds they'll recall the press conference as his announcement of his homosexuality rather than his emphatic denial of it."

In other words, as Ann intuitively recognized, the actor's mere mention of homosexuality increased the likelihood that people watching the press conference would later recall that he had called the conference to announce that he was gay. In short, he failed to retain control of the context and would have been better advised not to call the press conference at all.

In a paper titled "Credibility of Repeated Statements: Memory for Trivia," experimental psychologist F. T. Bacon shows that if a falsity is repeated frequently enough, we will begin to believe it. And Bacon isn't the only one who has confirmed this. As other researchers have elaborated, it doesn't matter whether the false information is presented as fact or opinion or whether it is written or spoken—it will be believed even when the information is identified as false from the very beginning.

"Warning people about false information also tends to make it more familiar later on," according to Ian Skurnik, one of the authors of a paper in the *Journal of Consumer Research* entitled "How Warnings About False Claims Become Recommendations." "People may correctly remember the information as false immediately after the warning. But as time passes, if people can't remember the details of the warning, but still remember the core piece of information, they will tend to think of it as true."

Could a similar dynamic prevail when it comes to historical events? If so, repeating untruths often enough will increase the likelihood that some people will believe them. An intuitive understanding of the power of this illusion-of-truth effect accounts, I believe, for the vigorous opposition from all parts of the world toward Holocaust denials. In many countries in Europe it's a crime to state publicly that the Holocaust never took place. Why such an extreme response to a patently preposterous claim, a claim that can be refuted by countless sources? Because the cumulative effect of repeated denials could conceivably lead to a "memory illusion." Our brain is organized in such a way that assertions, if repeated often enough, tend eventually to be accepted as facts. What is the basis for this so-called frequency-validity effect?

Familiarity is the key concept: the more something is repeated, the more familiar it becomes. If we hear something often enough, it becomes familiar to us, and we're more likely to believe that there must be something to it—an intellectual version of "if there's smoke, there's fire." And in many situations the linkage between familiarity and veracity holds true: The lion's share of what we hear or read repeatedly usually has some truth to it. In daily life patent falsehoods are usually eventually exposed and we discount the false assumption. Unless we're socially isolated or under undue influence from people who want to deceive us, the vast majority of our beliefs have some basis in reality. But despite this, most of us deep down are also aware of the perils that can emerge if false ideas remain unchallenged. That's because we recognize on some level that if false information goes unchallenged, we eventually become used to hearing it and thereby primed to accept it. Thus we're perpetually at risk for memory illusions.

Corrections of our first impressions of other people must also be carefully scrutinized. Despite claims about the power of "thinking without thinking," as touted by a recent popular book, our brains work best in evaluating new information when we slow up a bit and consciously attempt to discover whether our instant impressions really hold water. For instance, if I introduce you to a man who looks like your boss, you're going to unconsciously transfer to this new acquaintance many of your ideas about your boss. In fact, it's likely that you'll attribute many of the personality characteristics of your boss to this complete stranger. And this attribution will affect a whole array of your responses, including your general impressions, your mood, and your behavior toward that

→ self-fulfilling prophecy.
proj. ident. (as the stranger begins to behave or enact your projections)

person. Further, these responses will occur largely outside of your conscious awareness.

According to Susan Anderson, an experimental psychologist at New York University who has carried out much of the significant work on implicit impressions, "One comes to like or dislike the new person by virtue of some minimal resemblance he or she has to someone else—based on the overall feelings associated with that other person."

This transference effect can be even subtler. Suppose, for instance, I ask you to tell me some of the significant personality characteristics of your sister. I record your description and several weeks later I describe to you a new person (you never actually meet the person) who, from my description, shares one or more of the characteristics you described to me earlier as true of your sibling. Would that make any difference in your appraisal of that person?

Based on the results of an experiment Anderson carried out, you will be more likely to attribute additional characteristics of your sibling to this new person. In this case no physical comparisons are involved. I never actually introduced you to the new person; she may not even exist. I simply described her to you using the same or similar words that you employed to describe your sibling. Everything else took place unconsciously in your brain.

As a consequence of this approach to personality evaluation—thinking without thinking—you may embrace or reject someone for reasons that have nothing to do with their actual personality.

The Nefarious Power of Negativity

Unfortunately, our brain is more affected by negative than positive information. For instance, imagine these two scenarios. In the first you learn that you've won a $500 gift certificate from Saks. You would feel pretty good about that, wouldn't you? In the second scenario, you lose your wallet containing $500. How unhappy would you feel about that?

According to the results of risk-taking research, the intensities of your responses to these experiences differ markedly. As the result of what scientists refer to as the brain's negativity bias, the distress you're likely to experience as a result of the loss of $500 will greatly exceed the pleasure you feel at winning that gift certificate. Negative events, it turns out, mobilize elements of the fight-or-flight response: We feel anxious, our pulse increases, our breathing seems a bit more effortful, and so on. What's more, we're unpleasantly aware of these changes.

In addition, the brain needs additional time for processing negative information. To experience this for yourself, get someone to write down, using various ink colors, two lists of words describing desirable traits (*handsome, intelligent*) and undesirable traits (*untrustworthy, dishonest*). Ask them to come up with a list of at least twenty or thirty words in each category. When they're finished, ask them to time how long it takes you to read not the words but the ink colors the words on each list are written in. If you're like most people who have taken this test, a telling discrepancy will appear: It will take you longer to name the ink colors for those words describing undesirable traits compared to words for the desirable traits.

An automatic vigilance mechanism seems at play here: Negative events preferentially attract the brain's attention and exert a greater impact on brain processing than positive events. You probably won't have any difficulty coming up with everyday examples of this. Maybe you're looking forward to a holiday get-together but aren't wildly enthusiastic about listening to another long dinner-table political harangue by your brother-in-law. So which of these two subjects are you thinking the most about? It takes a deliberate effort to concentrate on the positive aspects of the holiday experience. Of course, it can be done, but why should it require effort? After all, your brother-in-law's speech couldn't possibly take up more than a few minutes (and may not even take place at all). So why do you keep mentally wrestling with it? Because the brain, if unchecked and left to its own natural rhythms, will always concentrate on risks, inconveniences, hassles.

Not all of us are equally prone to this negativity bias. We all know people who are pessimistic and given to bouts of depression or anxiety; we also can readily think of other people who are sociable, confident, and active. These two personality traits (referred to by psychologists as neurotic and extroverted personalities, respectively) typically respond very differently to emotionally arousing situations. To avoid having to deal with emotions, the neurotic typically avoids people; the extrovert, in contrast, is an inveterate glad-hander. The brains of neurotics and extroverts also respond differently to other people's emotions. When a neurotic looks at pictures of fearful faces, her amygdala reacts much more strongly than the brain of an extrovert looking at those same pictures. Smiling faces, in contrast, elicit a muted response in the neurotic's brain compared to the brain of the extrovert.

"Part of the reason extroverts seek social contact more often than neurotics may be that their reward system responds more positively to other people's smiles causing the extrovert to feel greater pleasure when they are around other people. On the other hand, individuals high on neuroticism may have brains that overreact to negative emotions, leading them to experience more anxiety and depression," according to Tal Yarkoni of Washington University in St. Louis, a student of the human brain's responses to emotions.

While this tendency to become ensnared in negativity varies from one person to another, it nonetheless remains true for all of us that, in the words of the researchers who carried out the word color test you did a moment ago, "negative information weighs more heavily on the brain and has a greater impact than equally extreme positive information."

Since negative information "weighs more heavily" on the brain, it's important to actively counter it with a more positive interpretation of events. This is not to say that we should deny painful, unpleasant, or awkward situations; rather, it's a suggestion that we should actively entertain alternative interpretations even under the worst of circumstances. While it's painful to be fired from a job or to be asked for a divorce, such experiences open the way for new jobs and new relationships. Of course, we can take this approach only so far: Some experiences (say, the death of a child) are so unsettling that it's difficult to envision anything positive about it. But even when dealing with extremes, our awareness of that negativity bias enables us to make active efforts to counteract it.

Take, for instance, racial stereotyping—probably the most serious consequence of this tendency of our brains to dwell on negativity. If you've had a bad experience with a member of a minority group—or even just heard or read something unfavorable about the group—it's only too easy to react prejudicially toward a member of that group. In many instances, the whole process may occur outside of conscious awareness.

In one of John Bargh's early experiments some of the participants were subliminally exposed to pictures of African Americans. Afterward all of the participants played a game in which the object is to help one's partner guess a word by providing clues. The game is challenging but sometimes frustrating, especially when one's partner seems oblivious to even the most obvious clues. During the game Bargh measured the hostility directed at these literally "clueless" participants. He found that those previously exposed to the subliminal pictures reacted with greater hostility than the other players. Even more interesting, they were more likely than nonprimed participants to experience their partners as being hostile to them—and with some justification, according to Bargh: "For the primed participants, their own hostile behavior, nonconsciously driven by the content of their stereotype of African Americans, caused their partners to respond in kind, but the primed participants had no clue as to their own role in producing that hostility."

Fortunately, we have a good deal of control over any racial stereotyping that we may be prone to—we're not at the mercy of our amygdala nor completely under the control of our unconscious perceptions. We know this on the basis of an intriguing experiment measuring racial prejudice.

The Prejudicial Brain

Reams of data exist on the subject of how you can spot a racist. For instance, the greater a white person's negative attitude toward blacks (as measured by a test of unconscious stereotyping that you can take for yourself by visiting https://implicit.harvard.edu/implicit), the greater the activity occurring in the person's amygdala when she looks at a picture of a black person compared to a white person. Since the amygdala is part of our brain's emotional circuitry and becomes active whenever a person feels uneasy, wary, or threatened, such a finding in a person with racist tendencies has been suggested as a useful means of distinguishing racists from people who are free of racial stereotypes. While such a claim sounds reasonable, I'm happy to report that things aren't quite so simple.

As it turns out, it matters a good deal how much time a person spends looking at the pictures of white and black faces, as well as the reason she's looking at the pictures in the first place. If the pictures are flashed on a screen so rapidly that she isn't consciously aware of them, the amygdala calls out its accusation: *racist!* But if the person is allowed more time to consciously look at the picture and formulate a response, the enhanced amygdalar activation doesn't take place. Instead, the firing occurs in the prefrontal areas of the brain, especially the anterior cingulate, a structure involved in higher levels of attention and the damping of those more primitive impulses originating in the amygdala. Why the difference?

In essence, the more thoughtful and tempered responses are acting to subdue the hurtful prejudicial ones. Although the first response may indicate racism, the response changes when the prefrontal lobes are allowed time to come into play. So is the volunteer subject a racist or not? It depends on how you want to define racism. While the initial response, over which the volunteer has no control, may indicate prejudice, the prejudicial response can be eliminated if that person is given the opportunity to use brain areas linked to rational reflection.

Perhaps you're thinking that the amygdala response is the only indicator that is really important, that the "real person" is the one reflected by that automatic activation of the amygdala. If that's how you think of it, you're not alone. That idea is commonly held in our culture. Typically, the argument goes something like this: Even though the "real" self may be momentarily "repressed" by more socially acceptable sentiments, it eventually emerges, oftentimes spontaneously and unbidden. According to this assumption, automatic responses reveal the "real" you, with the follow-up response entailing nothing more than an attempt to make nice. But when you think about it, why should those automatic responses over which we have no control be granted precedence over our more thoughtful reactions, which reflect our consciously espoused beliefs and values? In other words, is the real you accurately revealed on the basis of unbidden, transient, and unconscious thoughts and impulses that erupt over the span of milliseconds? Or is the real you the personality that emerges after your primitive automatic responses have been dealt with by your higher brain centers?

As a specific example, consider an interracial couple I once knew whose marriage fell apart because of a heated argument during which the white woman referred to her black husband with a racial epithet that flared up before her frontal lobes had time to damp it. Although she deeply regretted her outburst and immediately apologized, her husband never forgave her for this once-in-their-marriage outburst of racism; within a year they were divorced. So here's the crucial question: Was she "really" a racist who had successfully hidden her racism until the fateful moment of that instantaneous amygdala-driven outburst, or were the years of her generally satisfying marriage more accurately reflective of her real personality? Personally I would favor the latter choice if forced to decide about her real personality. At the very least, I'm suggesting caution whenever making a definitive statement about another person's real self.

Context is also important when deciding about the "real" us or the "real" them. In a thought-provoking study carried out by psychologist Susan Fiske, white participants were asked to look at pictures of black faces and mentally perform a series of tasks. In one task they had to sort the pictures according to perceived age (over or under twenty-one). This set up, according to Fiske, an "us-them" mind-set of "category-based emotional responses" that activated the amygdala. Next the volunteers looked at the pictures and decided the likelihood that the person in the photo might fancy a particular vegetable. In this task activation of the amygdala didn't take place. Why? Because this latter experiment forced the volunteers to consider the faces as belonging to unique individuals rather than to members of a group. Thus, according to Fiske, prejudice isn't immutable but rather socially constructed and modifiable. Change the context and you change the brain's response.

Black entertainer Sammy Davis Jr. understood this principle and applied it during the course of his fifty-year career to encourage the members of his audience to see him as a uniquely talented entertainer, rather than as a member of a racial group: "I always go onstage anticipating what people out there may be feeling against me emotionally. I want to rob them of what they're sitting out there thinking: Negro with all the accompanying clichés. Ever since I recognized what prejudice is, I've tried to fight it, and the only weapon I could use was my talent."

If stereotyping resulted only in prejudiced perceptions, the problem would be manageable—a prejudiced person would simply recognize his prejudicial tendencies and take extra care in preventing their outward expression. But, unfortunately, prejudicial perceptions may under certain circumstances lead to tragic behavioral consequences.

Over the last decade numerous incidents have occurred in various cities involving the shooting of unarmed suspects by police officers. Since this typically involved a white officer shooting a black suspect, the natural question arises: Was the officer's overresponse motivated by cultural and social stereotypes linking blacks with aggression and violence?

To find out, University of Colorado social neuroscientist Joshua Correll recruited college students to participate in a computer game simulation focusing on whether race influences a police officer's perception or misperception that a suspect is holding a weapon. In the simulated situations white and black actors held either guns or nonthreatening objects such as cell phones or wallets. In response, the participants were required to make speeded decisions to "shoot" armed targets by pressing one button and "not shoot" unarmed targets by pressing another.

A consistent pattern emerged: Participants were more likely to shoot unarmed blacks compared to whites. Further, their deadly force response also occurred faster for blacks compared to whites when both were carrying weapons. And when mistakes occurred (shooting an unarmed target or not shooting an armed one), electrical measurements of the brain's response to conflict showed differences consistent with racial stereotypes: Shooting someone not associated with violence generated the greatest conflict, but shooting someone stereotypically associated with violence wasn't any more of a problem than not shooting the person.

Although the team of neuroscientists who carried out the shooter video game readily admit that they don't know what transpired in those real-life situations in which cops shot unarmed black men, "our results suggest that the police officers' behavior was not as strongly regulated as it might have been if the victims were white," concluded the researchers.

Fortunately, few of us are ever placed in a position where we have to make an instantaneous decision and then immediately respond with deadly force. For most of us, whatever racial prejudices we may harbor play themselves out in time frames much longer than milliseconds. That increase in duration provides some measure of comfort: We are responsible not for whatever thoughts come into our heads over the space of an instant but only for how we subsequently act in response to these thoughts.

Admittedly, all of us differ in our unconscious attitudes toward other people based on variables over which we have little control: where we grew up, whether we regularly encountered minorities. Perhaps genetic factors may even be relevant here. But none of these factors is sufficient to label anyone as a racist. Rather, the critical factor is what a person *does* about these feelings. We're responsible for our actions and remain free to choose our responses. For reasons that aren't always clear, some people experience spontaneous and unbidden racist thoughts when encountering a member of a minority group. But their behavior isn't irrevocably determined by these thoughts; they can free themselves from them via deliberate efforts, as in the Fiske experiments.

"Unless one is socially isolated, it is not possible to avoid learning evaluations of social groups, just as it is not possible to avoid learning other types of general knowledge. Having acquired such knowledge, however, does not require its conscious endorsement," according to Elizabeth Phelps, lead author of the racial bias study mentioned above. "Both amygdala activation as well as behavioral responses of race bias are reflections of social learning within a specific culture at a particular moment in the history of the relations between social groups," says Phelps. "When people read about such brain image findings they pick up the mistaken message that there is a brain excuse for everything. That makes many people think it's not under their control, when in fact, it doesn't mean racial prejudice is any less controllable just because it shows up in an image taken of the brain."

Phelps's point is liberating and refreshing. We are not slaves to the automatic responses mediated by our amygdala and other components of our brain's emotional circuits; instead, thanks to the frontal lobes of our brain, we have the power to create for ourselves a new and more empowering reality.

A summary of what we've covered in this chapter seems in order. A few pages back we discussed how we recognize others we know by watching them move—even if we're watching their movement in terms of nothing more revealing than moving points of light in the dark. From here we looked at how one person's actions can directly influence another person's brain (that tea-drinking example) via the activation of mirror neurons. Next, we looked at how the MPFC is activated whenever we try to envision things from another person's point of view (the Sally-Ann test and the bank robbery story). We then examined how the brain falls prey to certain powerful propensities that appear to be hardwired in its circuitry: the illusion-of-truth effect, the power of negativity, an inborn tendency to react hostilely to people who differ from us. Notice how we have gradually segued from objective measures (recognizing another person from his or her gait) to subjective parameters (entering into another person's state of mind).

If social neuroscience has one basic tenet, it is this: Since we spend most of our time either thinking about or engaging with others, our brain's most important function is to free us from the prison of our own minds and enhance social communication. On most occasions this works best if our brain makes clear distinctions between ourselves and other people. But on other occasions such a strict demarcation between you and me leads to dehumanization and failures of empathy: Your feelings of grief, exhilaration, and so on remain totally inaccessible to me, and vice versa. In Chapter 5 let's take up how social neuroscience is revealing what is happening in the brain when one person experiences empathy for another.

5

The Empathic Brain:
Blurring the Boundaries
Between Self and Others

The Rubber Hand Illusion

Picture this: Across the room from you a man is sitting at a table with his right hand resting in his lap. In front of him on the table sits a lifelike rubber right hand exactly aligned with the real hand resting on his lap and hidden from view beneath the table. Using two small paintbrushes, a researcher is simultaneously stroking the rubber hand and the real hand. While this strange ritual is taking place the man is undergoing an fMRI.

After about 11 seconds the man begins to experience the stroking sensation as originating from the rubber hand instead of his real hand. The stronger this feeling, the greater the fMRI activity recorded from the premotor cortex. "Point to your right hand," the researcher then requests. The man points with his left hand to the location of the rubber hand instead of his real hand under the table. A moment later, Henrik Ehrsson, who carried out the rubber hand experiment, takes a hammer and pretends that he is about to smash the rubber hand. The man flinches, mentally makes an effort to move the rubber hand, and then expresses momentary surprise at his inability to control it.

The rubber hand illusion (a simple experiment that you can test for yourself with the help of a confederate) suggests that our awareness of our own body can be altered by manipulations involving our sense impressions. "The study shows that the brain distinguishes the self from the non-self by comparing information from the different senses," Ehrsson says. "You could argue that the bodily self is an illusion being constructed in the brain."

At the moment the illusion begins, the touch representations of the hand are altered: The sensations from the hand hidden beneath the table become tethered to the timing of the observable brush strokes delivered to the rubber hand on the table. It takes less than a quarter minute for the sense of ownership to be shifted from the real hand to the rubber hand. Thanks to this illusion, what is seen (the brush strokes against the rubber hand) takes precedence over what is felt (the brush strokes applied to the real hand beneath the table). As a result, the rubber hand comes to be felt as part of the body because, according to Ehrsson, "self-attribution depends on a match between the look and feel of the body part."

Ehrsson's experiment carries more profound implications than one might at first expect from such a simple experiment that is no more, really, than a kind of magic trick that can easily be conducted in a drawing room (minus the fMRI components, of course). Once the illusion arises, the "feeling" of ownership becomes remarkably powerful and persistent: Even though the man knows full well that the rubber hand isn't his, he still mistakenly points to it when requested rather than to the real hand. Although the exact mechanisms underlying his experiment remain speculative, Ehrsson believes that this "re-location of body-space," as he refers to it, involves the activation within the premotor cortex of "hand-centered cells," thus generating the feeling of ownership of the rubber hand.

As illustrated by Ehrsson's experiment, the boundaries and qualities of our sense of self are a good deal more malleable than we might predict. Empathy involves a similar blurring of the boundary between self and others.

At its most basic, empathy involves experiencing something that is happening to another person as if you were experiencing it yourself. In order to empathize, you have to be able to imaginatively establish a link between yourself and that other person. That's why we can empathize with a person but not with, say, a chair; chairs don't have feelings and don't experience the world in any way. Animals occupy a middle ground. They obviously aren't like chairs, but their inner experiences also aren't the same as our own (as far as we can determine). No doubt people with great empathy for animals would disagree with that last statement. They would certainly have a point when it comes to pain: An animal's response to pain closely resembles how humans react under the same circumstances.

But empathy involves sensitivity not only to the other person's experience but to our own as well. We can feel another's pain only if at some earlier point in our life we've experienced pain ourselves.

Consider this experiment: A group of volunteers is watching a man whose hand is strapped in a machine that allegedly generates a painful heat (a white lie, since no heat is involved). Some of the observers are told simply to make careful observations about the man; others are instructed to imagine how the man is feeling; still others should imagine how they would feel if they switched places with the man. People in the latter two groups who engage in deliberate acts of imagination show greater empathic responses as measured both physiologically (palm sweating and blood vessel constriction) and verbally (self-reports). Yet despite their empathic identification with the man, the participants in this experiment never lose their individual sense of identity. Unlike what happens in the rubber hand illusion, their sense of reality remains intact: The man receiving the heat shocks isn't perceived to be identical to the observer. Here empathy remains a controlled process in which the distinction between one person and another isn't altered. But if you change the above experiment slightly, a person no longer retains control over his or her empathic responses:

A woman and her boyfriend are together in a room, she lying inside an fMRI machine while he sits beside the machine. From inside the fMRI the woman can see her boyfriend's hand, which, along with her own hand, is hooked up to an electrode. The electrode periodically gives one or the other of them either a weak or strong one-second electric shock to the back of the hand. A message on a computer screen indicates to the woman whose hand will be zapped and how intense the experience is likely to be.

When the woman receives the shock, the fMRI shows activity across a wide swath of the brain, including the area where touch is perceived (the somatosensory cortex), along with areas responsible for the emotional experience of pain. When the woman isn't shocked but instead watches her partner being shocked, the activation pattern visibly changes: Her somatosensory cortex remains inactive, while the emotional centers once again light up like bonfires on a beach at night. Even though she isn't feeling her boyfriend's pain, she's experiencing the emotional components of that pain; the shock experience isn't happening to her, and yet at the same time it is. Sixteen different couples undergoing this test provided a vivid demonstration of the neurological underpinnings of empathy: The higher a woman's score on standard questionnaires gauging empathy, the greater the activity in the emotional areas of her brain.

When I first learned about the woman's response to the shock delivered to her boyfriend's hand I wondered what would happen if the same experiment were carried out with a sadist as the observer of the other person's pain. Would a sadist's brain respond by activating the same emotional circuitry? Although no such experiment has ever been done, I would expect that the findings would vary quite a bit from normal and involve the recruitment of the pleasure centers—corresponding with the pleasure that sadists derive from watching or inflicting pain on others. This gets us to a dark but I think important insight that is often neglected or denied in discussions about empathy: Empathy doesn't necessarily lead to positive consequences.

During a conversation twenty years ago the famed psychoanalyst Heinz Kohut mentioned to me that he believed that many of the Nazi interrogators were very empathic. I objected that empathy would prevent a person from hurting another. "Not if inflicting pain aroused pleasure in them," he responded. "Empathy involves getting into the mind and feelings of the other person—once you have that information, you can put it to good or bad uses," he concluded.

I thought of what Kohut had told me when I recently encountered a quote from a security service interrogator: "You have to know instinctively the exact time when to shout, when to speak loudly, when to speak quietly, or when not to speak at all and just sit and look at him—for hours if necessary. These things are instinctive. I have a thousand different systems for a thousand detainees."

The interrogator's instinct for how to proceed in a difficult interrogation stems from his empathic insights into how the prisoner will feel when subjected to a specific interrogation technique. Change the prisoner and the interrogator is able to select a successful technique from a "thousand different systems." What would turn up if the interrogator volunteered for a scan of his brain while he carried out an interrogation? I suspect the pleasure centers would illuminate like roadside flares as he applied his interrogation (some would say torture) techniques. But this is only a guess, of course. Although I'd be eager to see such an fMRI scan, I doubt that anyone engaged in interrogation techniques would submit to one. So, in the absence of an fMRI of an interrogator's brain, let's return to the couple who participated in the empathy experiment mentioned a moment ago in order to explore another aspect of empathy.

How the Brain Recognizes Faces

Imagine that several days later we're visiting that woman and man (let's call them Helen and Rob) from the empathy experiment described earlier. We find them in the midst of a squabble. Let's listen in:

> SHE: "I could tell you weren't in a good humor as soon as you came through the door."
> HE: "I didn't say a thing to indicate that."
> SHE: "You didn't have to. It was written all over your face! Now, instead of admitting that you were out of sorts, you want to deny it. Why don't you at least own up to your own feelings."

Helen and Rob are caught up in a common conflict situation: One person claims, against protest, a privileged access to another person's mind. In the fMRI experiment described earlier, Helen imagined what the shock would feel like and responded as if she were feeling the shock. Here she is doing something a good bit subtler: reading Rob's facial expression and drawing conclusions about his state of mind. Instead of detecting pain, Helen is detecting—or so she is claiming—Rob's emotional state.

Despite the subjectivity involved, perceptions such as Helen's often turn out to be true. That's because reading another person's emotions from his facial expression is an essential requirement for social survival that all of us learn at an early age. Autistic people lack this ability entirely; skilled negotiators and poker players rely upon it for their success; artists depend upon it for revealing the soul of their subjects. In Iris Murdoch's novel *The Sandcastle* the artist Rain Carter provides my favorite expression of the power of the artist to intuit the inner meaning behind a face:

> Painting a portrait is not just a matter of sitting down and painting what you see. Where the human face is concerned, we interpret what we see more immediately and more profoundly than with any other object. A person looks different when we know him—he may even look different as soon as we know one particular thing about him. . . . Perhaps we *feel* our own face, as a three-dimensional mass, from within—and when we try in a painting to realize what another face *is* we come back to the experience of our own.

Social neuroscientists now have some basis for explaining the observations of Iris Murdoch's fictional character: how one person can pull off the seemingly improbable feat of intuiting another person's state of mind from their facial expression.

First a little background. Facial recognition plays a particularly important role in our social behavior because the face is the preeminent means for recognizing oneself and others. If our ancestors weren't able to distinguish the faces of friends from enemies, you and I wouldn't be around to speculate about the importance of facial recognition. Of course, we don't just recognize people by looking at their faces; when answering a telephone we often recognize the caller by the sound of her voice. But for the most part, facial recognition is the commonest way for one person to recognize another. And since perceiving and identifying another person's face is so basic to social interaction, we are at a great disadvantage when we aren't certain who we are dealing with: We don't know how we should relate to the person, interpret his or her responses, or plan responses of our own.

"Human beings are alone among the animals in revealing their individuality in their faces. The mouth that speaks, the eyes that gaze, the skin that blushes, all are signs of freedom, character and judgment, and all give concrete expression to the uniqueness of the self within," as social critic Roger Scruton puts it.

As an indicator of the importance of the human face in social interaction, we come into the world already equipped to recognize faces. As infants we will look at a human face (usually our mother's, but any face will do) in preference to anything else that comes within sight. In adults the talent for recognizing faces depends upon a facial recognition center that exists on the underside of the temporal lobe in an area referred to as the fusiform face area. If that area is damaged, the recognition of faces is severely altered or even disappears altogether. (More about that in a moment.)

As you've probably noticed, the above paragraph presents us with a chicken-or-egg dilemma. Do our genes specify that a specific part of the brain should respond to faces? Or do our extensive experiences with faces over our lifetime lead to the development of a "face area"? The infant's predilection for faces at birth would seem to favor the first explanation, but the issue isn't completely resolved; infants may depend on imprinting or some other mechanism (the preference for faces disappears when toys, especially moving toys, capture the child's visual attention). So which comes first: a genetic predisposition for facial recognition or a kind of practice-makes-perfect arrangement whereby our facial recognition abilities get better with practice?

Here's a neat way of getting our answer. Assemble two groups of twins, one group (identical) sharing the same genes and a second group (fraternal) sharing about the same number of genes as shared by any two siblings. Now show both groups slides of human faces and record their brain responses by fMRI. When neuroscientists at the University of Michigan carried out this study, they came up with the answer to the chicken-or-egg question. Maps for facial recognition in identical twins looked very much alike while maps of the fraternal twins varied as much as you would expect when testing two siblings.

"Heredity plays a significant role in the pattern of cortical activity associated with face recognition," according to T. A. Polk, the lead scientist carrying out the study. "People are innately wired for the recognition of faces, perhaps because the ability to recognize faces is crucial to survival."

Although Polk is talking here about biological survival in jungles, deserts, and other inhospitable places, our survival has more social overtones. If you fail to recognize your supervisor's wife when you encounter her in a department store, that pleasant conversation you had with her at last year's holiday party may not be of much help in salvaging your job. (One wonders how many perceived snubs are actually the inadvertent results of the other person's underdeveloped facial recognition abilities.)

Considering the social importance of facial recognition, it should come as no surprise to learn that distinct types of facial recognition failures may occur, usually as a result of brain injuries. Such injuries result in devastating disruptions in both self-recognition and empathy.

Morphing Marilyn Monroe

Imagine yourself waking up one morning and encountering a stranger looking out at you from the bathroom mirror. In alarm you flee to the kitchen only to encounter another stranger who claims to be your spouse. When she speaks you recognize that she's telling the truth: Her voice sounds right, but her face isn't recognizable. You know that she's your wife when she speaks, but when she remains silent she doesn't look like anyone you know.

Unfortunately, this nightmarish situation isn't from a story by Franz Kafka but only too real for a small number of stroke victims. In these instances, the stroke (usually in parts of the right and left occipital lobes) results in what's referred to as prosopagnosia (from the Greek prosop-, "face," a-, "not," and gnosis, "knowledge"). The person afflicted with prosopagnosia, or face blindness, cannot recognize people from their faces but must rely on other sensory pathways, usually hearing the sound of the person's voice. Sometimes this nightmarish condition improves; sometimes it lasts for a lifetime.

At the other extreme from prosopagnosia, some of us are virtuosos in facial recognition, instantaneously identifying people we haven't seen in years. Others of us recognize people when seeing them but often can't put a name to the face: "Your face is familiar but your name escapes me." Still others can't recognize a familiar face when encountering that person in unaccustomed surroundings (they become flummoxed when meeting someone in a supermarket whom they met for the first time last summer on a beach). On average, though, most of us do pretty well on facial recognition. Typically, people remember up to ten thousand faces and can identify 90 percent of their classmates thirty-five years after leaving school.

Rather than being evenly distributed between the two hemispheres, facial recognition ability resides principally in the right hemisphere. We know this on the basis of a clever experiment done with computers involving gradual transformations of face pictures of different celebrities.

In one study, a research team headed by Julian P. Keenan anesthetized a volunteer's right hemisphere via an injection into the artery serving that hemisphere. Keenan then showed the volunteer a picture of herself "morphed" by computer with a picture of Marilyn Monroe. The resulting picture was a mix of the features of the subject and the movie star. When asked to identify the picture, she failed to perceive any of her own features but confidently identified the movie star as the subject of the picture. Anesthetizing her left hemisphere, in contrast, didn't interfere with the woman's ability to recognize her own face in the morphed picture. Here we encounter a temporary alteration in the brain's capacity for facial recognition in a normal person brought on by anesthetic.

In another morphing study, the face of Marilyn Monroe (she seems to be a favorite among brain researchers) was gradually morphed into that of former British prime minister Margaret Thatcher. In contrast to the Keenan experiment, no injections were given and nothing was done to modify normal brain function. For the most part viewers didn't detect the morphing process. Instead of perceiving the elements of one face gradually merging into the picture of the other face, the participants in the experiment tended to hang on to their original identification. Thus a picture consisting of 60 percent Marilyn Monroe and 40 percent Margaret Thatcher was identified as an older version of Marilyn Monroe; a 40 percent Marilyn and 60 percent Margaret mix resulted in the perception of a "sexier" side of the former prime minister. Eventually, as the morphing increased even more, the perception suddenly flipped: The "older" Marilyn was now identified as Margaret; the "sexier" Margaret Thatcher was finally identified as Marilyn Monroe.

Three different brain areas became active while the participants made their decisions. A pair of structures at the back of the brain, the inferior occipital gyri (IOG), concentrated on analyzing separate physical characteristics (lines on the face, makeup, etc.);

the second, the right fusiform gyrus (RFG), distinguished between the faces, probably by comparing the face in the picture to recognizable ones. And finally, if the face was recognizable, the third area, the anterior temporal cortex (ATC), which stores memories about specific people, identified the picture when the face was well known (that is, there would be a greater ATC activation to Margaret Thatcher among Brits compared to people from other countries).

According to neuroscientist Ralph Adolphs, such separate brain areas, each dedicated to specialized functions, correspond to *modules:* anatomical or functional brain components that are specialized to process certain kinds of information. Such an arrangement makes possible, says Adolphs, "an interconnected system of regions" responsible for how we see and identify faces. And when one or another of these modules breaks down, different failures in facial recognition emerge. A failure in the IOG results in an inability to pick up on important identifying characteristics such as the wrinkles signifying middle age. People with damage to the ATC find it hard to link a name with a face. Damage to the RFG results in the mistaken belief that different faces belong to the same person.

Not only is our social brain predisposed to turn its attention to faces, but also on many occasions another person's facial expression may be exerting an unconscious influence on us.

How Our Brain Resonates to the Emotions of Others: Emotional Contagion

In a paper entitled "Unconscious Facial Reactions to Emotional Facial Expressions," the authors found that when we look at another person's facial expression we tend to unconsciously adapt the same expression on our own face, even though we aren't consciously aware that we're doing so. This is called emotional contagion. For example, while watching someone smile, we activate at a subthreshold level the same facial muscles. This often has the effect of making us smile back. Why? Because perceiving an emotion activates the same brain circuits used to generate that emotion. Thus we're not only smiling in response to that person's smile but also experiencing a similar feeling of happiness. That doesn't always happen, of course. If a stranger smiles at us under certain circumstances we may inhibit any tendency to smile in return lest our smile be misunderstood. Even among friends we'll hold back the smile if we wish to communicate to them that we're not feeling all that happy at the moment.

Aside from the few exceptions just noted, emotions are infectious. We spontaneously respond to how other people are feeling. Indeed, just sitting in a room with someone can be enough to cause us to "catch" their mood, according to a classic experiment carried out a quarter century ago. Highly expressive people are especially capable, without saying a word, of passing along their negative moods to others. All of us to a greater or lesser extent resonate to the emotions of those around us.

On occasion, one person may be exposed to an emotion in another person that he or she doesn't enjoy empathically experiencing. For instance, as a physician, I experience this emotional resonance every day. Merely entering the waiting room and encountering certain depressed patients can set off a wave of depressive feelings in me, especially on days when I'm not feeling all that up to start with. Fortunately for me, I've been trained to recognize and not give in to this emotional contagion. But I think unrecognized emotional resonance may play a large part in the explaining the high suicide rate among doctors, especially psychiatrists. It also explains why physicians sometimes take seemingly unreasonable dislikes to certain patients. These negative countertransference reactions (to use the technical term for them) arise most often when the physician is attempting to ward off the negative emotions of a patient. Under such circumstances the doctor's impatience, anger, tactlessness, and other hostile responses (which patients rightly resent) are serving as protective reflexes against being engulfed by a patient's negative emotions.

You don't have to be a doctor to experience such feelings. We all know people whom we would prefer to avoid if given our choice. It may be because they always bring up gloomy topics or go on at length about how hopeless everything is. If we talk to them often enough or long enough, we begin to pick up on their negative vibes. So in order to avoid those vibes, we avoid the people.

But let's return to emotional resonance as it applies to Helen and Rob, who are engaged in their argument. Even though Rob doesn't recognize his own ill humor, for Helen it's as obvious as, well, the expression on his face. In response to Rob's facial expression, she momentarily experiences an attenuated version of the same negative emotions felt by Rob. As can be deduced from her comments, she's not at all happy about that. In response, she takes the offensive and confronts Rob, who in turn feels ill-used. After all, he hasn't actually *said* anything that provides objective proof that he's in a bad humor.

If Helen wants support for the correctness of her intuition about Rob, she need look no further than the philosopher David Hume. In his *Treatise of Human Nature* Hume speaks of an inborn "sympathy" or "propensity" that most of us have to pick up on the "inclinations and sentiments of others however different from or even contrary to our own." He observed that our minds are mirrors that reflect others' passions and sentiments.

The Empathy of Infants

David Hume's inborn "sympathy" for picking up on the sentiments of others correlates nicely with "intersubjective sympathy," a quality first described by the Scottish psychologist Colwyn Trevarthen that can be traced to infancy.

As noted earlier, newborns routinely pay greater attention to the human face than to anything else around them. Not only that, they also imitate facial movements performed by the adults caring for them. Mouth opening, lip pursing, tongue protrusion—if the infant sees any of these facial movements performed by the mother, he or she will do the same. Interestingly, the influence works both ways. When feeding their babies, mothers often open their mouths, and not always as an inducement to get their infants to eat since, in many cases, their mouth opening occurs *after* the infant has already done so. (The mothers are usually unaware of this sequence, incidentally.) And if the emotions attending the feeding situation are focused on, the results are even more interesting.

In a cruel experiment that makes me wince every time I watch the film depicting it, a mother deliberately stops facially responding to her infant by miming a neutral face during the feeding. Instantly, the infant turns fussy and looks away. It's as if a delicate and complicated dance between mother and infant has been disrupted. The existence of this behavioral dance between infant and mother has led behavioral researchers to suggest that from birth our brains are hardwired to resonate with other brains—whenever we observe another person's behavior, a mental representation of that behavior occurs in our own brain.

"Early imitation shows that interpersonal bonds are established at the very onset of our life (as early as 18 hours), when no subjective representation can yet be entertained because at that early point in life the baby isn't conscious," according to Vittorio Gallese, a neuroscientist at the University of Parma, Italy.

By the third month of life the infant's ability to automatically imitate disappears, later to be replaced by a more mature type of imitation. Both types of imitation rely on a similar process, according to Gallese. "What is common between an infant who replies to his mother sticking out her tongue with an equivalent behavior, and the skilled repetition by an adolescent of the piano chords as demonstrated by the piano teacher? Simply this: If I want to reproduce the behavior of someone else, no matter how complex it is, or whether I understand it or not, I always need to translate my external perspective of that person into my own personal body perspective. This problem can be overcome if both the other person's actions and my imitation share a basic format within the brain. This is exactly the case."

When we observe other people, we are exposed to a full range of expressive power, which is not confined to their actions but also encompasses the emotions and feelings they display, according to Gallese. "When this occurs an emotional interpersonal link is automatically established. Empathy is the means by which this link is established."

So here we have the origins of the empathic response. Starting hours after our birth, we're able not only to mimic another's actions but also to respond in a manner that would be described as "emotional" in an adult: The baby turns fussy when the mother breaks the imitative dialogue. A similar response occurs in us adults; if you want confirmation, listen to someone's description of how they felt when snubbed.

"Given the ubiquitous need to belong and be accepted, and the consequences that arise if this need is not met, mimicry is important in routine, everyday life," according to Tanya Chartrand of Duke University. "Mimicry binds and bonds people. Over time its usefulness has shifted from physical survival to a stronger emphasis on social survival."

The Chameleon Effect

Few of us go about our day consciously mimicking the actions of other people. Yet even though most of us aren't aware of doing it, we tend to unconsciously mimic a wide range of behaviors such as other people's accents, tones of voice, rate of speech, mannerisms, and even moods. In this process one brain is influencing another brain, often in the absence of any intention to do so. Mimicry doesn't even require another person's actual presence. Viewers mimic the facial expressions of people seen on television. We know this because of an ingenious experiment where participants watched one of two interviews: a person describing either a happy experience such as a surprise birthday party or a sad experience such as attending the funeral of a beloved relative. Unbeknownst to the participants, their facial expressions were being videotaped. In general their facial expressions matched the experiences being described in the interviews—happy expressions while watching the happy segments, sad expressions while watching the sad segments.

After a few days of exposure to a regional accent most of us find ourselves taking on some of the qualities of the local dialect. Such unconscious mimicry serves a useful purpose: It establishes a connection with the other person, fosters empathy, and serves as a kind of compliment (unless, of course, it crosses over into caricature). Mimicry also elicits more positive personality characteristics. In an intriguing experiment, people who had been mimicked proved more helpful and generous toward a wide range of people, not just the person mimicking them.

Psychotherapists and counselors have long been aware of the

power of mimicry. Ask any skilled therapist and he or she will tell you that the surest way of creating a positive rapport with a patient or client is to subtly mimic their gestures and mannerisms. I'm reminded here of the 1960s therapy guru Carl Rogers, whose much-caricatured technique involved reflecting back the patient's last comment ("Feeling bad, you say?" / "Yeah, Doc, really bad." / "Really bad?" and so on).

If you look at some old films of Rogers at work, you'll notice that more is involved in his technique than just endlessly reflecting back the client's previous comment. He's also mimicking many of the client's movements and gestures. But Rogers was careful not to go too far and make the mirroring too obvious, lest the client catch on to what he was doing and thereby realize that his or her sense of being understood wasn't real—that any perceived rapport was based on a gimmick. Skilled therapists are also aware they will lessen the effectiveness of their therapy by "anti-mimicking"— doing just the opposite of what the patient is doing.

Anti-mimicking exerts a similar effect outside of the psychotherapist's consulting room. If you're sitting comfortably but erectly in your chair, you're likely to react negatively toward anyone sitting across from you who slouches; if you're feeling a bit fidgety and moving around a lot, you're not going to respond well to someone nearby sitting rigidly and making few body movements. Nor does it seem to make a lot of difference whether you know the person well or he or she is a comparative stranger, according to Tanya Chartrand, who experimentally confirmed "the chameleon-like nature of unconscious mimicry."

"Like a chameleon that changes its colors to blend or fit in with its environment," she says, "people unwittingly change their mannerisms and behaviors to blend and fit in with their social environment. Participants in our experiments even mimicked unlikable people, indicating that even under minimal conditions in which there is no rapport, affiliation, or liking, nonconscious behavioral mimicry will occur."

In general, however, the longer people have known each other, the more likely it is that mimicry will have a formative influence on their behavior. And the more we like a person, the more likely it is that we will mimic them. Researcher M. LaFrance discovered this in his study of the interaction of teachers and students. Not only do students tend to mimic their teacher's posture, but this postural mimicking is greatest for those teachers the students feel the most rapport with.

Especially common is the mimicking of facial expressions between people who have known each other a long time. In a study of married couples, those married twenty-five years or longer were judged as being more physically similar than newlyweds or unmarried couples chosen at random. It's speculated by the experimenters that the longer-married couples have had more time to mimic each other, thereby developing the same facial lines and expressions.

Unconscious mimicry of favored people even extends to politics. In an experiment designed to measure emotional reactions to political leaders' expressive displays, the researchers concentrated on different people's emotional reactions to a taped news conference by Ronald Reagan. While Republicans reported feeling happy or angry in sync with Reagan's expression of these emotions, the Democrats in the audience showed no tendency to mimic or even respond to Reagan's emotions. Thus the natural tendency to mimic can be overridden, it seems, by a generally negative appraisal of a particular person.

Edgar Allan Poe's Experiment

If you really want to experience someone else's feelings at any given moment, especially in an emotionally charged situation, try this empathy-inducing exercise: Turn away for a moment and mime that person's facial expression. You'll thereby induce in yourself the inner experience corresponding to the emotion the other person is experiencing. You can do the same thing at a later point by miming the person's expression in a mirror. If you make yourself look angry, disgusted, afraid, sad, or surprised, not only will you inwardly experience subtle elements of these emotions, but specific and reliable changes will also take place in your autonomic nervous system, such as alterations in heart rate, finger temperature, and galvanic skin response (a measure of electrical conductance).

Edgar Allan Poe described how the miming of other people's emotions can lead to the inward creation of similar emotions in oneself in his short story "The Purloined Letter": "When I wish to find out how wise, or how stupid, or how good, or how wicked is anyone, or what are his thoughts at the moment, I fashion the expression of my face, as accurately as possible, in accordance with the expression of his, and then wait to see what thoughts or sentiments arise in my mind or heart, as if to match or correspond with the expression."

In line with Poe's insight, empathic individuals unconsciously mimic the postures, facial expressions, and mannerisms of other people to a greater extent than nonempathic individuals (the chameleon effect). This suggests that in order to empathize with another person we have to invoke within ourselves a kind of inner blueprint or representation of the bodily signs that characterize the emotions we are witnessing in that person. Method actors intuitively grasp the essentials of this. If an actor wants to fine-tune his depiction of, say, how a father feels when he's informed that his son has been killed in combat, the actor will want to know what the father did with his hands when he first heard the news, whether he stepped backward or forward, and, most of all, what changes took place on his face. Did he blink, close his eyes for an instant, or open them wider and stare off as if into an infinite space?

What occurs in the mind of the actor when he or she, as part of a role, attempts to arouse a specific emotion by using a previous life experience? Actress Ellen Burstyn described the process for Jonathan Cott in his book *On the Sea of Memory:*

> Let's say I had a situation in a play where I had to experi-
> ence grief. If I approach it directly, trying to remember
> some time when I felt grief, the emotion usually retreats.
> But if I approach it through my senses and I, say, picture
> the clothes I was wearing then and see if I can feel the feel
> of the clothes on my body with my fingertips and then try
> to remember what room I was in and where the light was
> coming from, and where the window was, see if I can feel
> the light on my face, and . . . go through all of the senses . . .
> then as I create all of those various sense memories, the
> emotional memory will follow.

Until recently it wasn't possible to scientifically corroborate Poe's insight nor the intuitive convictions of Method actors such as Ellen Burstyn. But two intriguing experiments reveal what's happening in the brain during the arousal of emotional memory.

In the first, Marco Iacoboni of UCLA asked volunteers to either simply observe a series of facial expressions or actively imitate them. The imitators, but not the observers, showed a boost of activity in emotional centers such as the amygdala.

A second experiment on empathy carried out in 2003 showed

that imitation and observation of other people's emotions activate a largely similar network of brain areas. To review (and oversimplify slightly), the brain contains two well-defined and separate systems. The first is concerned with emotions and consists of the limbic circuitry; the second is concerned with the construction of a blueprint of observed movements and involves a network composed of parts of the frontal, parietal, and temporal areas, especially the superior temporal cortex (STC). When a monkey stares at another monkey grasping a peanut, that action is encoded within the second monkey's brain in the form of a blueprint for performing the same action, as we mentioned earlier in our discussion of mirror neurons. That blueprint involves the same cells that the observing monkey would actually use if it imitated the first monkey and reached out for its own peanut.

Whenever you observe or imitate the facial expressions typical of the various emotions, alterations occur in areas associated with that emotion, as well as parts of the premotor cortex that activate the facial movements used to express that emotion. "We ground our empathic resonance towards others in the experience of our acting body and the emotions associated with specific movements," according to Laurie Carr, who led the team that carried out the experiment on empathy I just described. "To empathize, we need to invoke the representation [blueprint] of the actions associated with the emotions we are witnessing."

As Iacoboni puts it, "The way we understand the emotions of other people is by simulating in our brain the same activity we have when we experience those emotions."

Iacoboni's point is similar to Poe's suggestion in "The Purloined Letter," mentioned above. Poe, it turns out, anticipated suggestions such as that of Frans de Waal, an expert on empathy: "If you want to become more empathetic, you have to try to look at how people act and move their body and their face. Try to mimic it a bit, and you will feel internally what other people feel."

One caveat should be placed on de Waal's comments, it seems to me. Most of us are predisposed, perhaps even genetically hardwired, to feel the greatest empathy—and act on those feelings—with people who are similar to us. Instances of this arise every day. For instance, during the period immediately following the Southeast Asian tsunami in December 2004, millions of people watched televised videos depicting in grim detail the ocean sweeping people away, the anguished faces of parents separated from their children, and the steadily mounting numbers of dead bodies wrapped in body bags or simply sewn into sheets or whatever other material was at hand. Such scenes aroused general feelings of concern, but the pace of humanitarian relief for tsunami victims increased dramatically in Western countries after reports began surfacing concerning thousands of missing people from Australia, Germany, Finland, Norway, Sweden, and the United States who had been in the area on vacation. (These estimates turned out to be excessive and eventually revised downward.) Once viewers could more closely identify with the tsunami victims, contributions to the disaster relief efforts increased dramatically. That's because, according to de Waal, humans are inclined to empathize with others, but only up to a point: The most intense empathy—leading to decisions to act on the basis of that empathy—requires some sense of closeness, even identity with the victim.

"No animal can afford to feel pity for all living things all the time," writes de Waal in his book *Our Inner Ape*. "This applies equally to humans. Our evolutionary background makes it hard to identify with outsiders. We've been designed to hate our enemies, to ignore people we barely know, and to distrust anybody who doesn't look like us. Even if we are largely cooperative within our communities, we become almost a different animal in our treatment of strangers."

This observation is neither an expression of cynicism nor a criticism. A moment's self-reflection on one's own empathic sense—if carried out objectively and without preconceptions—reveals the not-terribly-flattering insight that for the majority of us empathy is most easily aroused in regard to people whom we perceive as most like ourselves.

Wouldn't it be nice to be able to fine-tune and enhance our capacity for empathy so as to include people with whom we seem to have little in common? I'm thinking here of achieving the level of empathy spoken about by Christ ("Do unto others as you would have them do unto you") and other religious leaders and prophets throughout history. Neuroscience can be helpful here and is now making a major contribution toward understanding what for a better term we might call the deeply empathic brain.

Compassion and the Frontal Lobes

At the W. M. Keck Laboratory for Biological Imaging at the University of Wisconsin, neuroscientist Richard Davidson has been studying over the past few years what happens in the brains of experienced practitioners of the ancient practice of meditation. One particular form of meditation is especially relevant to our goal of understanding empathy.

Loving-kindness-compassion meditation calls for the practitioner to think about a loved one and then generate feelings of love or compassion for this person. For example, she might imagine the loved one in a sad situation and then mentally desire for that person freedom from suffering and the achievement of well-being. After finishing this exercise, the practitioner generates these same loving and compassionate feelings toward all living beings, without thinking of anyone in particular. While the meditation practitioner engaged in these imaginative exercises, Davidson and his team carried out fMRI recordings of the practitioner's brain.

Earlier in his career Davidson demonstrated that the frontal lobe (specifically, the prefrontal cortex) plays a prominent role in the experience of both positive and negative emotions. He's found the left frontal lobe to be more involved in positive emotions, the right frontal lobe more in negative emotions. As a result, damage to the left frontal lobe by a stroke often results in depression (due to the loss of positive emotions ordinarily generated by that lobe); right frontal lobe damage, in contrast, may result in just the opposite—instead of depression (an appropriate response to brain damage) the patient shows a blasé, couldn't-care-less response that may progress in some cases to outright denial that the stroke has even occurred.

During loving-kindness-compassion meditation, Davidson found an increase in activity in the left frontal lobe for all subjects—a finding consistent with the association of that area with positive emotions. Activation also occurred in a network comprising deeper-lying areas of the brain known to be associated with movement, the observation of another person in pain, and maternal love.

"These findings support the view that our experience of another's suffering is mediated by brain regions involved in our own experience of pain," says Davidson. "Thus the etymology of compassion, meaning to 'suffer with,' makes a great deal of sense. So too does the expression 'moved by compassion.' "

I thought of Davidson's findings when I first learned that Buddhist monks in Thailand were playing a major role in the grim task of cremating thousands of victims of the December 26, 2004, tsunami. According to several monks involved in the cremations, they were able to carry out this stressful work without developing post-traumatic stress disorder (PTSD) thanks to their many years of practicing a special form of meditation known as "corpse meditation." In this seemingly bizarre practice the meditator concentrates on a picture of a dead body—sometimes a horribly decomposed one—in order, in the words of one monk, "to remind us of the true nature of life, which is its impermanence and transitory nature."

Although I'm not an expert on Buddhism, I recall a point made frequently by the Dalai Lama: Impermanence and compassion are interlinked. It's easier to arouse compassion toward the suffering of others by thinking about the transitoriness of their lives. "Compassion is not simply the wish to see another being free from suffering, it is strengthened by insight into the transient nature of existence, such as the realization that the being who is the object of your compassion does not exist permanently," according to the Dalai Lama.

Now, if you equate compassion with empathy, then these seemingly unrelated items add up to an empowering and transcending concept: We have a greater ability than we've previously imagined to become more empathic by inducing positive changes in our brain.

According to Davidson, "Loving-kindness-compassion is mediated by brain functions involved in positive emotions and maternal love and empathy." But that's not nearly as exciting as the conclusion Davidson has reached on the basis of his research: "Achieving this brain state is a skill that can be learned."

Not only will learning this skill be helpful to the normal brain, but it can also be therapeutic for people with mild forms of brain irregularity. For example, a child (or adult) with attention-deficit/hyperactivity disorder (ADHD) may benefit from a form of meditation practice known as Shamata, according to another study carried out by Davidson and other neuroscientists at the Keck Laboratory, University of Wisconsin.

Shamata meditation involves focusing all of one's attention upon a single object such as a small dot on a computer screen. The goal is to keep the mind focused entirely on the dot while ignoring anything going on either in the environment or within one's own head (wandering thoughts, memories, etc.). If one becomes momentarily distracted by outer perceptions or inner thoughts, the remedy is to refocus attention on the dot.

Two groups participated in the study: Shamata meditators with ten thousand to fifty thousand hours of meditation practice, and a control group composed of a group of people with no previous training in meditation. Both groups were scanned in an fMRI while they stared at the computer dot.

Although both groups used the same brain regions during the Shamata meditation, the activation was much stronger among the experienced meditators. In addition, the brain activation of the inexperienced meditators slowly decreased in strength over the course of three minutes of meditation.

"The experienced meditators showed more brain activation as they gradually gained stability in the meditation, while control subjects started out in a focused state but gradually became more and more distracted or sleepy," according to Julie Brefczynski-Lewis, of the Keck Laboratory.

Which areas of the brain were more active in experienced meditators? Two main areas showed selective activation. The first was the visual cortex, responsible for representing the visual dot that served as the object of meditation. With lapses in attention— such as thinking about a favorite song or an item from a to-do list—activation in this area decreased. During such distractions, attentional resources are diverted from the visual cortex to regions that represent the song or the to-do list.

Second, the frontal-parietal cortex also showed greater activation in the experienced meditators. The frontal areas (near the top and front of the head) and parietal areas (near the top and back of the head) are known to be active during attention-demanding situations. Damage to these areas leads to profound attentional neglect: A person may fail to pay any attention at all to one side of visual space, thus bumping into objects on that side. Such observations among frontal-parietal brain-damaged patients have led to the insight that the frontal-parietal network is responsible for the control of attention.

Experienced meditators have achieved superior control and stability of attentional focus, according to the Wisconsin study. But even more interesting is that the amateurs and experienced meditators activated similar brain patterns. The difference is that among the experienced meditators the activation remained steady, while among the inexperienced meditators the activation decreased in strength over the course of three minutes of meditation.

"The experienced practitioners have greater control and stability of emotional focus," according to Brefczynski-Lewis. "However, similar brain activation patterns in the two groups show that the control subjects were indeed correctly doing the meditation and had achieved some degree of proficiency and understanding of the practice. The results from this study give promising evidence that Shamata meditation is an effective way to train attention with results that one can see in brain scans."

Interestingly, the Wisconsin study provides scientific evidence for something Eastern thinkers have always insisted upon: The mind can be trained to overcome shortcomings by practicing and concentrating on areas of weakness. If attention is a problem, as with people afflicted with ADHD, than some degree of improvement can be expected by carrying out mind-focusing and mind-concentrating exercises.

If emotions are the problem, the Wisconsin study suggests a solution based on the following rule: In general, the emotions that we bring to an encounter with another person will be the same emotions that that person will reflect back to us. If you need convincing on this point, here's a little experiment you can carry out with the cooperation of an unattached male friend. Give him the name and telephone number of a woman you know. Tell him that she's beautiful and, in addition, she has a great personality. A day or so later, give him another name, but this time tell him she's on the plain side but, hey, a date's a date. Later talk to these two women after they've spoken with the man. You'll get descriptions that match up in an interesting way, according to the authors of a paper ("Social Perception and Interpersonal Behavior: On the Self-fulfilling Nature of Social Stereotypes") who carried out just such an experiment.

Men who had been led to believe they were talking with an attractive woman went out of their way to sound friendly and warm, which in turn elicited a similar response from the woman. But if the men believed that their prospective date was homely and not very exciting, they approached the conversation more matter-of-factly. The women, in turn, spoke less during the conversation and, in general, responded lukewarmly to the halfhearted offer of a date. In essence, the woman seemed to "catch" the emotion projected to them by the caller.

Not only do we all resonate to the emotions of people around us, but also in some instances we may even look to others for clues to how we should be interpreting our own feelings ("You're okay; how am I?"). For instance, if I inject you with a few drops of adrenaline, your heart is going to beat faster, your breathing rate will pick up, and you'll feel generally activated. Yet if I ask you what emotion you're experiencing, your answer will depend very much upon your surroundings, especially the behavior of those you're interacting with.

If they act giddy and silly, the chances are you will, too. If they are acting angry and resentful, you're going to feel out of sorts as well—at least that's the findings of an experiment involving the injection of adrenaline aimed at testing how we label our emotions. Even though everybody in the experiment was experiencing the same physiological arousal caused by adrenaline, the participants labeled their emotions in a way that was consistent with those around them (reminiscent of the effects of amphetamine on monkeys described in Chapter 1—a result that provided one of the original inspirations for social neuroscience).

Such findings lead, I believe, to a code of conduct that we should all adhere to. It's good mental hygiene for ourselves as well as a service to others to try to keep our thoughts and emotions as reasonably positive as we can, given the circumstances of the moment. So when heading off to an evening with friends, leave your melancholy mood at home.

6

The Power of the Frontal Lobes

Thinking Makes It So

Most of us can readily bring to mind from our early years examples of parents, teachers, or others urging us to "put on a happy face" or "look for the silver lining" in every unhappy experience. Sometimes such admonitions helped us feel better; sometimes they merely annoyed us. Eventually, most of us settled on a compromise: We recognized that, if nothing else, putting a positive spin on bad situations spares us the pain and attendant misery of making those situations even worse by immersing ourselves in dark thoughts about them. So it came as a pleasant surprise to me to learn from neuroscientist John Gabrieli about an exercise he carried out with some of his colleagues at Stanford and Columbia universities that provided scientific proof for the power of positive thinking.

You can participate in Gabrieli's exercise by staring for a few moments at a picture of a sick patient lying in a hospital bed. Nobody likes to think of themselves as sick in a hospital, but for a few minutes stare at that picture and do just that: Imagine yourself in the role of that patient. Really enter into the spirit of the exercise—rev things up to the point you can almost vicariously experience the sights and sounds surrounding that bed along with the patient's agonizing pain. Further, imagine that the person in the picture has been suffering for a long time and isn't likely to recover. While you're doing that, Gabrieli and his team are measuring your brain's activity. Here's what they will find, based on a Gabrieli exercise in which volunteers emotionally identified with the sick person in the picture.

As you become more emotionally involved with the events in the picture, a functional MRI of your brain shows increased activity in your left amygdala. As mentioned earlier, the amygdala is a key structure for processing fear, anxiety, and other negatively arousing emotions. The left amygdala is especially important in the mental representation of anything that elicits fear. Accompanying this increase in amygdalar activity, Gabrieli also finds greater activation in a part of the prefrontal cortex on the left side of the brain that is important in the retrieval of how it might feel to be in pain and confined to a hospital bed.

Now make a deliberate effort to reinterpret the scene in a more positive way. Imagine that the sick person isn't really so sick and that the pain isn't so bad; imagine that the person in the bed is more fatigued and tired than uncomfortable, has a strong constitution, and is on the way to recovery. As you perform this mental reinterpretation Gabrieli will record a different change in your brain. The amygdalae on *both* sides become less active, while part of your frontal cortex becomes more active.

"What we're seeing here is the effect on the brain of reappraisal," Gabrieli told me. "And reappraisal is something we do every day whenever we're faced with an emotionally disturbing or stressful situation."

Reappraisal can make situations seem either better or worse: We can remind ourselves that the traffic accident we encounter on our way to work may not be as serious as we imagine it to be as we drive by. On the other side of the equation, putting a negative spin on events isn't always undesirable (athletes deliberately cultivate aggression during a game, for example). The point here is that reappraisal involves transforming an experience by either emphasizing the positive aspects of that experience or placing the emphasis on the negative aspects. Whenever we do either of these things, we change our brain: Thinking makes it so, at least in terms of the changes we bring about within our brains.

"This strategy of cognitive reappraisal is based on the idea that what makes us emotional is not the situation we are in, but the way we think about the situation," notes Kevin Ochsner, a collaborator with Gabrieli. "Using cognitive reappraisal either to increase or decrease one's emotional responses engages [left] prefrontal brain regions that, in turn, either increase or decrease responses in the limbic system."

Our emotional responses can also be lessened if we identify and label our emotions, according to research carried out by UCLA neuroscientist Golnaz Tabibnia. For example, looking at pictures of angry faces activates the amygdala and elicits an emotional response. But the intensity of this response can be decreased, Tabibnia discovered, if the person mentally labels the face as "angry." Labeling involves linguistic processing that activates the prefrontal areas, leading to a decreased response from the amygdala. According to Tabibnia, "The prefrontal cortex attenuates responses in the brain's emotion centers. That's why emotion-labeling may help reduce emotional responses in the long term."

Fortunately, most of us most of the time can label and reappraise emotionally distressing situations, thereby depriving them of their sting. But, as is all too evident in everyday life, people differ in their ability to regulate their emotions. Some people are very good at it, whereas others perform miserably.

But however skilled or unskilled a person may be in regulating his emotions, age plays a part in determining the brain pathways involved. We have Mario Beauregard, a neuroscanner at the University of Montreal, to thank for this insight. In a two-part experiment he compared the fMRIs of women twenty to thirty years old to the fMRIs of girls ages eight to ten while they watched a series of sad film excerpts. Beauregard asked the women and girls to allow themselves to experience sadness in response to viewing the excerpts. Regardless of age, the same structures became active: the ventrolateral prefrontal cortex and the anterior temporal pole, brain regions known to be associated with experiencing sadness in healthy adults.

In the second part of the experiment, Beauregard asked the subjects to voluntarily suppress their sad feelings. The young girls had greater difficulty complying than did the adult women. Their fMRIs also differed. While both the girls and the women activated the same circuit comprising different parts of the prefrontal cortex (the orbitofrontal and the dorsolateral areas) and the anterior cingulate, only the girls activated the hypothalamus—a pivotal brain structure involved in the visceral ("gut reaction") dimension of emotion.

"These findings suggest that the less efficient emotion regulatory capacity seen in children may result from the fact that, in humans, the development of the prefrontal cortex is not complete until early adulthood," comments Beauragard. "But with the maturing of the prefrontal cortex, we humans have the capacity to consciously and voluntarily modulate the electrical and chemical functioning of our brains by voluntarily changing the nature of our mind processes." Nor is that capacity a small matter. "The ability to modulate emotions is at the heart of the human experience; a defect in this ability may have disastrous social and emotional consequences."

It's the prefrontal cortex that plays a major role in helping us to modulate our emotions. (Think back to Jonathan Meaden, whom we met in Chapter 1.) In order to further illustrate just how important that frontal lobe actually is, let's move from PG- and even R-rated research to some research bearing an X rating.

Now that I have your attention, let me modify that last sentence slightly: The research isn't X-rated. Rather, the subject matter of the research involved pornography and how the brain responds differently according to the attitude of the person encountering it. Beauregard enlisted the help of ten healthy men, ages twenty to forty-two, to watch excerpts of pornographic films while he measured their brain activity under two conditions.

In the first, they were told to allow themselves to become sexually aroused. Later they watched the films under the instruction to control their urges and act the role of "detached observers." Any guesses about the findings? Well, as you probably anticipated, the older, deeper portions of the brain such as the amygdala and other parts of the limbic system were most active during the arousal portion of the exam, while the men's prefrontal cortex assumed command when they tried to control their impulses and act the role of neutral observers.

In summary, activation of the prefrontal lobes reins in the unbridled instincts of the emotional circuits. And since the prefrontal lobes are among the last brain areas to mature, it takes time before sexual and other instinctual urges can be completely controlled. That's why "raging hormones" present such a difficulty for adolescents. Only it's not hormones that are to blame but, rather, the immature prefrontal cortex, which isn't as yet completely capable of controlling the adolescent's impulses. Of course, this insight was recognized long before Beauregard's research.

For example, the Catholic Church requires that a child must reach the "age of reason" before becoming responsible for his or her acts. If you substitute "prefrontal maturation" for "age of reason," you have a pretty good description of what distinguishes an adolescent and a young adult. Our legal system makes a similar distinction: Adolescents are remanded to detention centers for behavior that if committed by an adult might well result in a lengthy imprisonment or even worse. The controlling influence of the mature prefrontal cortex over the emotional circuits also explains what happens if the prefrontal cortex becomes damaged or diseased in some way: It's difficult to suppress instinctual urges and emotions.

But let's move from adolescents, pornography, and damaged brains back to normal adults, specifically Kevin Ochsner's observation, mentioned earlier, that some people are less skilled than others at self-regulating their emotions. Who hasn't on occasion mentally revisited for hours or even days an embarrassing or emotionally painful past event? But, as Ochsner observes, with some of us this tendency to ruminate (continue to think about negative events long after the upsetting situation has passed) takes on the quality of an obsession. What is the effect of a ruminative style on brain function?

If a ruminator is placed in an fMRI and asked to simply look at photos of emotionally arousing events, that person's brain will show a stronger amygdalar response compared to someone who doesn't mentally "chew the cud" (the derivation of the word *ruminate*) over negative situations. The amygdalar response is even greater when the ruminator doesn't simply look at the pictures but negatively reappraises them. And when asked to think more positively about the situation in the picture, ruminators show a weaker response than would normally be expected. In short, ruminators pay a price for their gloomy approach to life.

"Those who chronically ruminate about negative events in their lives or aspects of themselves are doing extra brain processing that may be increasing brain activations in deep emotion-related brain structures such as the amygdala. This makes it much more difficult for these individuals to regulate their negative emotions than it is for others who are not prone to rumination," according to Gabrieli. One of the solutions to this dilemma is to reestablish the balance between our emotional and our reasoning centers.

Specifically, we need to allow our prefrontal lobes to exert cognitive control over the emotional responses to emotionally arousing situations triggered by our amygdalae. We should aim at allowing our prefrontal lobes to provide alternative representations of events—balancing our emotions with less emotionally disturbing interpretations.

As an example of what occurs in the brain, imagine yourself looking at a picture of a woman crying outside a church. Think of the woman as attending a funeral of a close relative. Her grief has so overcome her that she has broken down into tears, seemingly unable to control her grief in front of other people. Clear your mind for an instant and now imagine her as a member of a wedding party and that she is literally crying tears of joy. If you later look at fMRIs taken during these two imaginative exercises, you will observe that activation occurs in your amygdala and parts of the orbital cortex while you are thinking of the woman as attending a funeral, but when you reappraise the situation, changing your interpretation of the picture from grief to joy, the amygdala becomes less active while other parts of the frontal lobes increase in activity. In essence, your simple reappraisal of a distressful situation brought about a change in brain function that, in turn, lessened your distress.

Nor do we ever lose this ability to enlist the prefrontal lobes to help us create positive interpretations. Take the "grumpy old man" phenomena. Despite the common conception that older people are more prone to see the dark side of things, additional research carried out by Gabrieli proves just the opposite. Possibly as part of the emotional wisdom that comes with aging, older people show reduced amygdala responses when shown depressing pictures. To this extent older people are *more* positive than younger people. Depressed people of any age, in contrast, almost can't stop themselves from putting negative interpretations on everyday events. Overall, studies such as Gabrieli's may soon make it possible to identify individual differences in people, whatever their age, to experience and control negative emotions.

Attachment Styles

Why do some people experience little difficulty mentally reappraising an emotionally charged situation while others, seemingly unable to alter troublesome thoughts, simply suffer? The difference seems to be related to what psychologists refer to as "attachment styles." People with *avoidant* attachment styles tend to be critical, aloof, and generally uncomfortable in intimate relationships. People with *anxious* attachment styles, in contrast, are preoccupied with closeness and tend to be suspicious that their affectionate feelings may not be reciprocated. In general, anxiously attached people experience great difficulty "letting go" of or suppressing negative thoughts, especially thoughts concerning rejection and loss. Avoidantly attached people, in contrast, are capable of suppressing or altering such thoughts fairly easily.

People with different attachment styles also differ in how their brains process disturbing thoughts when they arise. The brain of the avoidant cohort shows greater activity in prefrontal regions associated with reward and motivation. This pattern correlates with the ability to suppress disturbing thoughts. The anxiously attached person, in contrast, activates brain regions associated with stress and emotional processing—findings consistent with heightened anxiety. This activation of emotional circuitry explains why anxiously attached people experience such difficulty changing their negative thoughts by reappraisal.

But which comes first? Does the emotional circuitry of the anxiously attached person become active in response to disturbing thoughts, leading to difficulty in reappraising a distressful situation by enlisting the help of the prefrontal cortex? Or is that emotional circuitry already in action before the distressing thought even occurs, such activation even favoring the emergence of negative interpretations? So far, social neuroscience hasn't an answer to that chicken-and-egg question. Some of us may be predisposed to anxious or avoidant attachment patterns not because that's how we've chosen to view the world but because that's how our brains are organized. The hopeful element in all this is that, based on research findings by Gabrieli, Ochsner, and others, we now know that we can exert control even under the most difficult of circumstances thanks to our powers of reappraisal.

Take, for example, the placebo response. Since doctors are now ethically barred from using placebos, consider this situation instead: You go to your doctor complaining of pain. After a thorough evaluation she determines that nothing serious is going on. She prescribes a mild painkiller and several hours later you're feeling better. "That medicine really worked," you're likely to conclude. But suppose you later find out that the pharmacist had mistakenly given you a mild diuretic (fluid pill) that slightly increased your urinary output but should have had no effect on your pain.

Before you conclude "It was all in my mind," consider this: If a person believes that a pill lessens pain even when it isn't chemically designed to do so, it's likely that this mistaken belief decreases activation of the amygdala and pain-related structures such as the cingulate and the insula while increasing activation of the lateral and medial prefrontal regions, which are related to mental control. In short, your belief about the power of the drug prescribed by your doctor but mistakenly dispensed by the pharmacist activated changes in parts of your brain involved in the appreciation of pain along with areas concerned with reappraisal.

Negative reappraisal also exerts a powerful effect. If you are anticipating that a harmless medical procedure will produce pain, parts of your cingulate and temporal regions—areas involved in the generation of your emotions and expectations about pain—will activate. But change your expectations—get them more in line with what is realistically likely to happen—and you can dampen that cingulate-temporal activation.

Not only can we change the activation of our brains, but we can also alter our brain's chemistry, according to a research study involving professional actors. On request, seven healthy male actors self-induced one of three states (sadness, happiness, or a neutral emotional state) and maintained that state for the duration of a PET scan study (about sixty minutes).

During the actors' self-induction of the requested emotion, the PET scan measured serotonin synthesis capacity (SSC) in different brain areas. (SSC is a measure of the brain's ability to make the neurotransmitter serotonin from its chemical precursor, the amino acid tryptophan.) If a specific brain area makes a lot of serotonin, it will "light up" on the PET scan; areas making lesser amounts of serotonin will show no increase in activity. The study revealed both in the cortex and deeper in the brain regional differences in the SSC patterns associated with self-induced sadness versus happiness. This suggests different patterns of brain SSC corresponding to distinct self-induced emotions.

"We found that healthy individuals are capable of consciously and voluntarily modulating SSC by transiently altering their emotional state," according to Elisabeth Perreau-Linck, a brain scientist from the University of Montreal, who carried out the study. "In essence, people have the capacity to affect the electrochemical dynamics of their brains by changing the nature of their mind process. This is a kind of 'positive emotion therapy' that anyone can use to modify the chemical functioning of their brain."

Both the actor study and the mental reappraisal research illustrate the powerful effects we can exert on our own brain function by reappraisal. The Roman emperor Marcus Aurelius, who presumably knew little about the brain, anticipated this insight years ago: "If you are distressed by anything external, the pain is not due to the thing itself, but to your estimate of it; and this you have the power to revoke at any moment."

Change your thoughts and you change your brain; change your brain and you change your feelings. Remember this the next time you find yourself dwelling on the darker side of events in your life. You might not be able to change all of the distressing events around you, but you can change your attitude about them and thereby enhance your brain function and mental health at the same time.

The Brain's Response to Rejection: Why It Hurts to Be in Love

In order to learn more about the role of the amygdala in modulation of emotions I spoke with Naomi Eisenberg, an expert on what happens in the brain during social rejection, and specifically why some of us are so sensitive to social rejection. As she put it in a draft paper "Why It Hurts to Be Left Out," social connection is "as basic as air, water, or food" and "the pain of social separation or social rejection is not very different from physical pain."

In support of her argument she points out that our language is filled with metaphors linking physical pain with broken social bonds. We can be hurt by a dog's bite, or by an acquaintance's biting remark; we experience "hurt feelings," even a "broken heart." Such metaphors suggest, according to Eisenberg, that on a deep level social pain and physical pain share common features and common anatomical landmarks. Most prominent is the anterior cingulate cortex (ACC), the large structure that curves along the medial wall of the frontal lobe.

A person writhing in unbearable pain can be freed of the worst aspect of his pain if a surgeon makes an incision into the ACC. Following this operation (called a cingulotomy) the patient continues to feel the pain, but it no longer bothers him. It's possible, in other words, for a person to distinguish between a perception and his reaction to that perception. Eisenberg uses a metaphor of her own to explain this distinction: "Rating the intensity of pain can be likened to rating the loudness of the volume on a radio; rating the unpleasantness of pain is more like rating the extent to which you perceive an increase in volume as bothersome." While other brain regions process the intensity of the pain, the ACC is responsible for rating the unpleasantness of physical pain. And it isn't even necessary for actual pain to be involved. Just the anticipation of a painful experience can activate the ACC.

Imagine yourself in Eisenberg's lab, where you are involved in one of her experiments testing your response to pain. You are told that you will receive either a nonpainful touch or a painful shock to your skin. Your brain will process these two situations in different ways. If you know for certain that the stimulus will be nonpainful, your ACC remains quiescent. But if you're left in doubt as to whether you will receive a light touch or a shock, your ACC comes into play. As a result, even the nonpainful touch will be experienced as unpleasant. In other words, the mere possibility of pain heightens the sensitivity of the ACC. A similar activation of the ACC occurs under conditions of real or threatened social isolation. We know this based on research on creatures located a few branches away from us on the evolutionary tree.

Destruction of the ACC in infant monkeys will render them incapable of emitting distress calls in the absence of the mother monkey. The social dysfunction increases even more if the ACC ablation (a euphemism for surgically cutting the structure) is carried out on the mother monkey. She will no longer respond to her infant's distress calls (not surprisingly, the survival rate of the infants of such mothers falls drastically). But the mischief isn't confined to mother and infant. Such a surgically modified mother will also spend less time engaging in social interactions with other monkeys. It's as if she no longer cares one way or the other about the social networking that prevails in your typical monkey colony.

Since most of us don't live in monkey colonies, one can be forgiven for wondering about the relevance of this research to our own lives. And since ablating human ACCs isn't something that can be routinely carried out on normal people, the proof for the relevance of this research must involve subtler experiments.

Here's one I find both convincing and charming: Put a new mother in an fMRI machine and measure her brain responses as she listens to the sound of her baby crying. The ACC fires up for the baby's crying (the sounds of babies other than her own will do just as well). In contrast, not much happens in her ACC when the experimenter replaces the sound of baby crying with general background noise. Maternal responses, in other words, are hardwired within the brain.

Notice the progression in our understanding of the ACC. Starting with physical pain, we've segued into a discussion of social pain—the pain that accompanies abandonment or rejection. Naomi Eisenberg thinks that this continuum is a natural consequence of the piggybacking of social attachment to the pain system in the interest of promoting survival. "Social pain and physical pain share the same underlying system and this overlap has several consequences for the way these types of pain are detected, experienced and overcome," she says.

To prove her point Eisenberg presented during her lecture a series of slides showing people staring out at the audience with facial expressions of extreme disapproval. And although actors were only miming the facial expressions, these expressions were sufficiently discomfiting that the audience giggled nervously at the faces looking out at them with expressions of contempt or cruel humor or total exclusion. She then showed fMRIs taken of the brains of volunteers while looking at the slides. The images confirmed that the social rejection conveyed by these faces activates the same brain involved in pain sensitivity: the dorsal anterior cingulate cortex (dACC). Moreover, a person's sensitivity to this form of social exclusion correlated directly with their pain sensitivity.

The most painful rejection for most of us is when someone we've in love with decides that he or she is no longer in love with us. Not all of us respond to rejection in the same way. Most people, after a brief period of mental anguish that varies from person to person, move on and look to establish alternative romantic relationships. Others have a harder time letting go and brood for months, occasionally even developing a clinical depression that requires psychotherapy or antidepressant medication. Still others begin acting in ways that everyone recognizes as disturbed: stalking or threatening the former lover, sometimes even acting violently.

To further understand the brain mechanisms of romantic rejection I called on Helen Fisher, an ebullient and enthusiastic anthroplogist and neuroscientist from Rutgers University who has done research on the emotions associated with romantic love following rejection. As Fisher mentioned with a laugh at the beginning of our conversation, she had recently been dumped herself, "so I can personally identify with the people in my study." Combining a pragmatic view of romantic love ("It's involuntary, difficult to control, and generally impermanent") with a scientist's eye for the telling detail, Helen Fisher is that rare combination of scientist and humanist—just the ticket for correlating brain activity with different people's responses to rejection.

"When lovers are rejected, they generally protest and try to win the beloved back; eventually, however, they slip into a state of resignation and despair." In order to measure this, Fisher used fMRI to study the brain activity of eleven women and six men who had recently been rejected by their romantic partner. While in the fMRI each looked at photos of their former lovers.

Fisher found increased activation in the same brain areas associated with the reward system as occur in people who are happily in love. This isn't surprising, Fisher says, "since adversity tends to heighten romantic passion." But in addition, the brains of rejected lovers showed activation in different parts of the reward system (the nucleus accumbens and the ventral striatum, specifically) that light up when a person is involved in a high-risk investment or, perhaps more important, during cocaine craving.

"Brain systems associated with reward and motivation remain active in abandoned people but differ in their precise location. While looking at the pictures of their former lovers, brain areas associated with addiction become active. This may explain why some spurned men and women stalk a departing mate and why others become deeply depressed and even commit suicide or homicide when the partnership fails. In these instances, rejected men and women continue to feel attachment for their abandoning partner, as well as craving and romantic love," she says.

Fisher's findings are certainly consistent with the behavior of people who don't seem to get it when a romance has ended. Although it is clear that the romance is over, the spurned lover keeps acting as though the sparks of that now defunct relationship can be rekindled. This initially puzzling behavior is less puzzling when viewed from the vantage of what's happening in the brain. Based on the brain activation patterns observed by Fisher, romantic love can best be understood as a kind of addiction (marked by craving, withdrawal, and the potential for violence)—hence Fisher's comments that romantic love is involuntary, difficult to control, and usually impermanent. Her findings are also right on target with the subjective feelings of the rejected lover as described by rocker Gene Pitney in his song "It Hurts to Be in Love": "It hurts to be in love, when the only one you love / Turns out to be someone who's not in love with you."

7

How Our Brain Determines
Our Moral Choices

The Sweet Emotions of the Soul

In search of outward clues to people's internal emotional states neuroscientists have learned a lot about reading facial expressions and other body language. In the process they have learned the importance of the brain in determining morality and ethics. This insight can be dated to Charles Darwin, who wrote in 1872: "The movements of expression in the face and the body reveal the thoughts and intentions of others more truly than do words, which may be falsified."

Even more influential than Darwin was Guillaume Duchenne, an eccentric and socially awkward French neurologist who during the middle years of the nineteenth century developed an obsession with applying an electrical impulse to the skin of people's faces in order to stimulate contraction of the underlying facial muscles.

In 1862 Duchenne published his findings in *The Mechanisms of Human Facial Expression*, a book that you're not likely to find except on the shelves of a collector of medical memorabilia. During research for his book, Duchenne, who was a pioneer in medical photography, took many pictures of his favorite subject, a thin, toothless, rather rumpled-looking figure referred to simply as "the old man." Since he was afflicted with facial anesthesia, the man was an ideal subject for Duchenne's experiments involving the application of painful electric currents to shock the facial muscles into activity. In some of the pictures the old man is laughing; in others he expresses surprise, sadness, or fear—the specific emotion depending on which of the facial muscles Duchenne was electrically stimulating at the moment.

"The sweet emotions of the soul," wrote Duchenne, activate the muscles around the eyes, whereas false or even halfhearted smiles involve only the muscles of the mouth.

Without any greater effort than looking up from this book you can train yourself to distinguish true from false smiles. For example, suppose you're reading this book on an airplane (if you're not, just think back to any of the unmemorable flights you've probably taken). Observe the airline attendant as he or she moves smoothly and effortlessly down the aisle, smiling indiscriminately at each customer. Does that smile make you any happier that you've decided to fly on this airline rather than a competitor? "Well . . . not really," you're likely to respond. "Flight attendants just smile like that at all of the passengers because they're paid to."

Indeed, such a tepid false smile—long referred to by psychologists as the "Pan American smile" (presumably named after a flight attendant of that now defunct airline)—is likely to involve just the lower part of her face. But now suppose that as the flight attendant serves you your coffee you make a witty comment or briefly tell a really funny joke. If you've successfully amused the flight attendant, you'll notice the muscles around his or her eyes go into action, briefly crumpling the overlying skin into tiny furrows resembling crow's feet.

Paul Ekman, along with his longtime colleague Wallace Friesen, both from the University of California, have in the spirit of Duchenne aimed at melding the art and science of reading faces. The Facial Action Coding System (FACS) developed by Ekman and Friesen enables an observer to monitor all the facial muscles, decode the facial expressions, and correlate the expressions with the number of facial muscles in play. These observations provide a measure of how strongly and for how long the muscles remain contracted.

By using the magic of slow-motion photography in place of Duchenne's electrical prods, Ekman has confirmed the distinction between genuine and false smiles. If the muscles around the eyes don't activate, the smile is only for show. In tribute to his predecessor, Ekman has dubbed the full smile the Duchenne smile. And there are other differences in smiles: False smiles are less symmetrical than joyful smiles, remain on the face a microsecond too long, and tend to end abruptly rather than fade smoothly away.

Not surprisingly, Ekman and others interested in the quality of smiles have pondered about the role deception plays in all of this. After all, the formal smile is inherently deceptive: It invites us to believe that the person smiling at us is experiencing an emotion that he or she isn't. So wouldn't it seem desirable for all of us to be intuitive experts at picking up fake smiles from those smiles that activate "the sweet emotions of the soul"?

Well, desirable or not, most of us perform poorly when it comes to distinguishing true from feigned emotions as expressed in the two varieties of smiles. (If we've undergone formal training with something like the FACS, we do slightly better.) If you consider this a failing, here's another way of thinking about it: Most smiles are perfectly harmless and serve as a kind of social lubricant that facilitates interaction between people who but for the smile might be at each other's throats. Isn't it preferable in the interest of maintaining harmonious social relations to put on a smile one doesn't feel rather than frown or fret or otherwise express our displeasure?

In most cases, an inability to distinguish real smiles from shams actually works to everybody's advantage.

Notice how we've gone from a discussion of smiles to one of deception. The surest way to *hide* one's feelings is with a smile. "Put on a smiling face" carries with it the suggestion that activating the muscles of the lips and cheeks can conceal true feelings. Perhaps that helps explain why we're uncomfortable around people who stare back at us with a fixed smile on their face. While research on smiles provides some help here, other factors are obviously in play: A person may put on a Pan American smile because she's nervous or preoccupied, not really engaging with the people and situation of the moment. Other brain research suggests more helpful avenues for measuring trust. Take tone of voice, for instance.

Lie Detection

Impatience, irritation, anger, condescension—feelings such as these are usually easily detectable, even when the speaker is doing his best to hide them. "I'm really sorry to hear about your accident" conveys regret. But if the words are spoken without a corresponding regret in the tone of voice, we may have reason to doubt the speaker's sincerity. On the other hand, when we're convinced of another person's truthfulness we say things like "He sounds sincere." That way we justify our belief that that person is telling us the truth and therefore we can trust him.

It's the right hemisphere of our brain that is specialized to make such determinations. A patient with damage in that hemisphere may lose her ability in an ambiguous situation to detect the emotional intent in another person's voice. As a consequence, she has a difficult time fathoming such phrases as "Way to go!" Is the speaker expressing admiration or withering sarcasm? To answer that question, even those of us who are mercifully free from right hemisphere damage must listen carefully to the tone of voice, extracting the intended meaning by detecting subtleties in the delivery of a particular phrase.

For most of us, on most occasions, the detection of dysprosody (abnormalities or irregularities in speech inflection, stress, and rhythm) comes naturally. Even so, we may be momentarily flummoxed if the person we're conversing with employs a deadpan delivery; at such times we must look for other clues such as gaze, facial expression, or bodily movement to reveal the speaker's intended meaning. But for our present purposes let's concentrate on tone of voice.

A voice conveying dominance is deep, loud, moderately fast, and clearly articulated. Granted, it's sometimes difficult to distinguish dominance (which is bad) from confidence (which is good). When in doubt, other people's responses give the first clue whether a person is coming across as confident or overly pushy. Confident tones inspire and motivate others; dominant tones incite resentment and anger. And this distinction carries practical consequences.

Surgeons who speak in dominant tones are more likely to be sued by their patients than those who speak with a touch of concern or even empathetic anxiety in their voice. Presumably, the patients can intuitively distinguish a dominant tone from a voice that expresses concern and empathy. And since the patients judge such speakers to be caring, they make less inviting targets for malpractice claims if something goes wrong in the treatment.

Listening to a person's tone of voice makes it possible to detect "leakage" information—lapses that occur when a person attempts to deceive others as to their true feelings or intentions. Indeed, the tone of voice is the most revealing of all sensory channels when it comes to detecting the speaker's true feelings or beliefs. So on those occasions when you want to catch another person in a lie, talk to him or her on a telephone rather than face-to-face. That's because most of us after a lifetime of practice possess highly honed skills for concealing facial or other bodily movements that might reveal aspects of ourselves that we'd prefer to keep hidden. But few of us (with the exception of actors) have developed an equivalent talent for controlling our voice.

Deliberate attempts to deceive another person ensnare the deceiver in an interesting paradox: the greater his motivation to appear honest, the greater his underlying anxiety, which leads to a higher vocal pitch and fewer head movements—a combination the insightful listener is likely to perceive as insincerity. That slight change in pitch resulting from underlying anxiety explains why successfully lying to the boss is harder to pull off than lying to a coworker. Hardest of all is lying to a spouse, who can draw on years of experience detecting tiny deviations in our vocal tones. But before anyone becomes too confident about detecting a spouse's deviation from the truth, another paradox must be kept in mind: When it comes to the detection of deception, the harder we try to detect a lie, the less successful we are at it.

"Lie detection is, for the most part, a nonconscious process," explains Nalini Ambady, associate professor of social psychology at Harvard. In support of her claim, Ambady cites a vast body of research suggesting that formal training in lie detection *decreases* the number of accurate determinations. In criminal investigations, for instance, those trained to spot liars are often less accurate than untrained people. But despite easily demonstrable limitations in their lie detection ability, those who have undergone formal training in lie detection usually express great confidence in their ability to spot liars. They also back up these false claims with a host of rationalizations in support of their mistaken judgments. But as Ambady discovered, a higher level of confidence doesn't ensure accuracy. Bottom line: Don't think that you're going to become a lie detection savant by taking a course in the "art" of deception detection.

"When left to their own devices people tend to use automatic strategies to detect untruthful statements. Thus not only are people largely unaware of their true lie detection abilities, but those who are more confident also show more false negative errors [i.e., failures to detect lying]," says Ambady.

Given these rather dismal facts about lie detection, it should come as no surprise to learn that brain scientists are on the lookout for a more reliable means of detecting liars. But that's not an easy task. For one thing, when it comes to telling lies the variations in frequency, style, and technique are staggering. At one extreme are the psychopathic liars for whom lying is second nature. At the other extreme are people who cannot tell even those minor deviations from the truth that are sometimes required in the interest of harmony and amicability. Such people can't seem to stop themselves from blurting out information that would be best concealed by means of a little white lie ("To tell you the truth, that dress really doesn't look good on you" instead of "You look marvelous").

For the most part, social neuroscientists have concentrated on how best to detect the virtuosos of lying who routinely escape detection (unless they come into contact with someone possessing highly developed intuitive abilities in picking up untruths). As I mentioned above, that kind of insight can't necessarily be learned. So that leaves us with the notion of coming up with something like a brain print for lying as especially appealing. But such a development is unlikely since a lie doesn't sit in a specific area of the brain awaiting an fMRI or other technique to unmask it. A lie is simply another piece of information—in this case false information.

But an fMRI may prove more successful in distinguishing a true memory from a false one. The distinction is important: False memories may lead to statements and claims that are untrue, and yet they aren't lies. For instance, I can create an illusory recollection in you by asking you to study a list of words containing such related terms as *thread, pin, haystack, sharp, point, pain, injection, eye, sewing.* When I present you a few moments later with a revised list containing the word *needle,* you're highly likely to falsely remember *needle* as being on the first list because of its association with all of the other needle-related words on the list. Even though *needle* didn't appear on the list, it will seem to you as if it had. This memory illusion, well established in the experimental psychology literature, is so powerful that you may persist in claiming that *needle* appeared on the original list unless I give you the opportunity to scan the list once again. (You might want to confirm this effect for yourself by carrying out this test on a friend.)

When the revised list is presented to individuals in an fMRI, the brain responds differently to a real word from the list compared to the falsely remembered word. The medial temporal lobes, regions associated with memory storage, contribute mainly to true memory formation. The left prefrontal cortex, an area associated with descriptions and memory elaboration, contributes to both true and false memories. This distinction between the two brain areas makes sense: Simple recognition takes place in the temporal lobes, while verbal elaboration ("These terms are associated with needles") takes place in prefrontal regions.

In short, true and false memories share some common mechanisms within the brain—one of the reasons false memories can sometimes seem so convincing even though they are false. And while an fMRI scan could conceivably be helpful in distinguishing a true memory from one that's false, it wouldn't be helpful in distinguishing a truth teller from a liar.

An imaging device is likely to demonstrate activation in the brain's emotional circuitry resulting from the conflict that results whenever a person knows one thing but says another. But monitoring the emotional circuits won't always be helpful, especially with people who don't experience a lot of emotional conflict when lying. To take the commonest example, no emotional perturbation occurs when the liar is convinced that the interrogator has no right to know something or has no right to be asking him questions in the first place. Thus issues of legitimacy enter into the truth-telling equation. If I'm convinced that a certain piece of information is none of your business, I may lie with perfect equanimity; further, no measurable alteration in my emotions will occur.

As an example of the problems involved in detecting a liar, consider a little white lie of my own (one that I suspect many of my readers can identify with). For my mother's eightieth birthday I planned a party with my brother and sister that included people dating back to various periods of my mother's life, about fifty people in all. Since we had decided on a surprise party, I faced certain challenges. First, I had to think of a reasonable excuse why my mother had to be out of the house the afternoon and early evening of the party so that things could be set up and the arriving guests greeted and made welcome. Even more difficult were the emotional challenges: coming up with a response when my mother asked why I hadn't so far made any more fuss about her birthday than handing her a card during breakfast, or why I hadn't invited my brother and sister to the house, or why I hadn't at least told her ahead of time about my lack of plans so that she could have spent the day elsewhere.

During this interrogation I experienced a panoply of emotions: eagerness to tell her about the party, disappointment that she really considered her eldest son capable of such insensitivity in response to such an important event, and annoyance at myself that I had agreed to a surprise party instead of insisting that we tell her about it ahead of time. To further complicate matters in this deception, I had to intercept several calls intended for my mother from people who had forgotten that she didn't know about the party.

What I remember most clearly about the experience was the sense of inner tension. Certain things could be talked about, others couldn't; reasons had to be formulated explaining why my mother shouldn't make certain calls ("I'm certain your sister hasn't forgotten your birthday. She'll probably be calling any minute now"); mail had to be quickly scanned to eliminate the possible delivery of cards or notes that might reveal the imminent birthday party.

If an fMRI had been taken of my brain during these few hours prior to the party, I expect it likely would have shown increased activity in the frontal lobes (responsible for formulating and sustaining the series of lies needed to keep everything secret), the temporal lobes on both sides of the brain (responsible for keeping my memory operating efficiently enough to balance the many elements required to maintain my deception), and the limbic system (responsible for the emotional roller coaster I was riding all that day).

Actually, my predictions about an fMRI taken at that time conform to the findings of Dr. Scott H. Faro, director of Temple University's Functional Brain Imaging Center. In a small study Faro placed several volunteers in a scanner and instructed them to lie about something they had recently done (firing a toy gun, of all things). Some of the volunteers hadn't fired the gun and served as controls. Since everybody answered no when asked if they had fired the gun, some of the respondents were telling the truth while the rest were lying. The fMRIs of the volunteers who were telling the truth were far less active than the fMRIs of the liars, which showed frontal, temporal, and limbic activity—the same pattern that I suspect would have marked an fMRI of me if taken while I was engaged in my deception.

"Since lying is a complex behavior there is not just one center in the brain," Faro explains. "Multiple areas are interacting.

There's more activity and more interactions that occur during a lie than in truth telling."

I take Faro's comments as supporting my own hunch that we're not likely to come up with a brain print for liars. There are just too many variables, too many factors that can't be controlled. (What about the person who isn't lying but giving out false information that he sincerely believes to be true?) And in order to activate the many brain areas associated with lying, a conflict must exist; but I can't be conflicted when I'm denying something that I don't re- member having done. And what about people who aren't lying but simply are uncertain of their facts? Finally, there are those who don't tell the truth because they truly believe that their inquisitor is violating certain commonly accepted boundaries. In such situa- tions it's likely that the amygdala, various components of the limbic system, and the frontal and temporal areas will shift into overdrive. But that isn't necessarily because the person is lying. He may sim- ply be momentarily uncertain about the best way to proceed.

Thanks to the complex interplay of factors involved in lying, it's highly unlikely that a single brain area or circuit will be discov- ered to be typical of all lies told by all liars.

The Runaway Trolley

But social neuroscience may have a bit more to contribute to our understanding of how we process moral dilemmas. Contrary to what we've been taught to believe, emotions rather than logic play the principal role in determining moral behavior. This is especially true in regard to the resolution of moral dilemmas.

In a moral dilemma a choice must be made among options that are all morally undesirable to a greater or lesser degree. For exam- ple, imagine yourself as a soldier in Iraq who encounters a suicide bomber headed toward a crowded marketplace while carrying a small child. As the bomber moves closer to the crowd, you have a choice: shoot him now in full awareness that you might also kill the child, or hesitate long enough for him to enter the marketplace and kill and injure perhaps hundreds of people. What would you do? Neither choice is morally desirable and would ordinarily set off internal conflict. But if that conflict goes on for more than a few seconds, the opportunity to act and thereby save many lives will be lost. So tell me—and quickly—what you would do.

Fortunately, most of us encounter less agonizing moral dilemmas, usually involving choices between what's in the long-term best interest of others versus what's most immediately appealing to us (i.e., donating a portion of an unexpected financial windfall to a hurricane relief effort versus using it to trade up to an even more expensive sports car). Or the dilemma may be more nuanced. During a business deal we must decide between long-term cooperation, which would result in higher profits for both parties, or deceiving the other party, which would allow us to grasp the immediate profits but would preclude any future transactions with that person.

Moral dilemmas involve a specific set of emotions (guilt, shame, embarrassment, and other states) called, fittingly enough, *moral emotions*. Because they involve real or potential interaction with other people, moral emotions arise later in life than the primary emotions (happiness, fear, anger, disgust, and sadness). This time sequence correlates with the brain's maturation: The primary emotions are present early in our lives and involve parts of the brain that are functioning at birth (areas beneath the cortex and in the brain stem), whereas the moral emotions require the participation of the frontal and prefrontal areas, which don't come online until later in life (starting in early adolescence in most people).

Thus the experience of moral emotions and the appreciation of moral dilemmas develop in tandem with the maturing of the prefrontal areas. And if those prefrontal areas never develop correctly or later become damaged because of disease (recall Jonathan Meaden, whom we met in Chapter 1), moral emotions may never develop at all.

Neuroscientist Joshua Greene has used fMRI to study what happens in the brain when people are faced with moral dilemmas. In his now classic hypothetical example, a trolley is bearing down on five people who will be killed unless somebody activates a switch that shuttles the trolley onto another track where a solitary worker is carrying out track repairs. Most people reluctantly choose to activate the switch and save five lives at the cost of one. But in the second part of the experiment Greene ups the ante by asking his subjects to imagine standing beside a stranger on a footbridge overlooking the tracks at a point between the trolley and the five people. All that's needed this time to save the lives of the five people is the willingness of the subject to push the stranger down onto the tracks below.

While the math is the same (one person dies, five people are saved), most people—and I'm sure you'll be as happy to learn this as I was—are unwilling to push another person to his death whatever benefits might accrue. Moreover, just thinking about pushing somebody from the footbridge activated different areas of the brain (mostly involving emotion and heightened vigilance) than the scenario that involved turning the switch. Most of the fMRI experiments on moral dilemmas carried out so far reveal increased activation of the very same areas that are critical for social behavior and perception (medial and orbital prefrontal cortex and the superior temporal area).

Nor should this activation of emotional circuitry come as a surprise. For the vast majority of us, the wrongfulness of pushing someone to his death involves strongly felt moral emotions aroused by what we perceive as a dastardly act. Because we are able to emotionally identify with the man on the footbridge, we react with revulsion at the thought of directly causing his death. And that revulsion exerts a more powerful effect on our conduct than reasoning about the wrongfulness of killing another human being. Brain imaging is thus providing support for the notion that ethics and morality aren't the products of reason acting alone. Indeed, reason and emotion aren't separate and distinguishable but work together in determining why we behave as we do.

Especially important for achieving moral ethical judgments is our old friend the frontal lobes (specifically, the orbital frontal cortex). Given the importance of that area in the formation of moral judgments, it should come as no surprise to learn that the frontal lobes are underactive in psychopaths and others who act without moral or ethical restraint. While psychopaths usually make the correct moral decision when presented in a test with a moral dilemma (they express an unwillingness to harm anyone), in a real-life situation they may act callously and brutally. Thus the psychopath when questioned says one thing ("I would help the old lady across the street by guiding her across the intersection") but when in the actual situation acts very differently (mugging and robbing her).

But what about people with greater-than-average moral and ethical sensibilities? Would the PET scan of such an individual, for instance, differ markedly from the average person's? No such studies have ever been done; but one thing seems certain: Researchers like Joshua Greene are now providing the pivotal insight that even though morality involves an inner subjectivity (one's conscience), social neuroscience can provide objective ways of studying it.

8

Make My Memory

The Road from Brain Research to Neuromarketing

Consider your memory for just a moment. If you're like most people, the further back in time you have to dredge, the more unreliable your memory becomes. But you may be objecting that there's one notable exception to this: "If something is sufficiently emotionally arousing, I'll remember it for years."

To prove your point you may mention various details about where you were and what you were doing on September 11, 2001, when terrorists carried out their deadly attacks on New York and Washington. "Just about everyone can tell you where they were and what they were doing when they first heard the news," you may insist. Sometimes referred to as "flashbulb memory," this heightened remembrance for emotionally arousing events is often considered a kind of gold standard of accuracy: If a person is sufficiently aroused, he will remember details about an event that would otherwise slip into the black hole of forgetfulness. If this is a rough précis of what you believe, it may come as a surprise to you to learn that flashbulb memories aren't especially reliable after all.

Heike Schmolck, a behavioral neurologist at Baylor College of Medicine, suspected as much and devised an experiment to test her suspicion. Three days after the verdict in the O. J. Simpson trial she asked people the details of their memory of the announcement of the verdict. She asked them again fifteen and thirty-two months later. After fifteen months, 50 percent of the memories matched the original descriptions very closely; only 11 percent contained major discrepancies. But during the second interview at thirty-two months, only 29 percent of the memories matched the original account, and 40 percent contained major discrepancies. For example, one college student who originally remembered hearing the verdict while watching television in a lounge with other students furnished an equally vivid flashbulb-like account thirty-two months later in which he remembered hearing the verdict while sitting at dinner with his family at home. Both memories were accompanied by a firm conviction of truth despite the fact that one of them had to be mistaken.

While writing this chapter I personally experienced the unreliability of an especially vivid memory. It surfaced the day Mark Felt was identified as the secret source for journalists Bob Woodward and Carl Bernstein, who broke the Watergate story and coauthored *All the President's Men*. This development led my wife and me to recall a party we attended a year prior to the Watergate break-in. Both of us remember the hostess coming over and telling us she wanted to introduce us to someone very interesting. From there our memories of what happened differ substantially.

I remember the party as taking place outside and the guests informally dressed except for the person our hostess introduced us to, Gordon Liddy, the subsequent ringleader of the Watergate break-in. I clearly remember him wearing a suit. My wife remembers the party as taking place indoors and that Liddy, in comparison to the other party attendees, was underdressed in a T-shirt and slacks. Since such a memory discrepancy was the very topic I was currently writing about for this book, I decided to call the hostess of that party in order to resolve our different recollections of the party. Needless to say, she had to think for a few minutes to recall a party she had given more than thirty years earlier. She then remembered the party in line with my wife's recollection rather than my own (an indoor party with Liddy informally dressed).

If I hadn't spoken to the hostess, I would have remained convinced that my memory was correct and my wife's recollection mistaken. But even though I am now persuaded that my memory was incorrect, I still cannot make the necessary modifications to bring my memory in line with what really happened. As with the student interviewed by Schmolck, I still see in my mind's eye that outdoor party in my own mistaken way. And although nothing terribly dramatic occurred that would necessarily qualify as a flashbulb memory (other than meeting someone who later became a notorious felon), I was absolutely convinced nevertheless that my erroneous memory was correct.

"Although we all have these memories and are very confident about them, they are nevertheless very often false; the longer the event in question has passed, the more likely it is that our brain has 'rearranged' the memory," comments Schmolck. "While we remember situations very well that went along with emotions, emotions don't help us memorize facts."

When I asked Richard McNally, a memory expert at Harvard, for an explanation about all this, he responded: "In many instances memory for the central gist of something that happened is retained, whereas memory for details fades or changes. People do not forget that the *Challenger* exploded or that terrorists destroyed the World Trade Center. But they may forget where they were and what they were doing when they heard about these shocking events, despite a subjective sense of memory's immutability."

Drawing on social neuroscience research showing the malleability of memory, a new breed of marketers is now busily devising practical applications of social neuroscience aimed at getting us to buy their products. But how can we protect ourselves from marketers and others who may seek to subtly alter our memories for their own purposes? For instance, do you believe that a picture may convince you of the reality of something that didn't happen? I learned the answer to that question firsthand one afternoon about a year ago via a practical joke my wife played on me. She casually called me over to the computer to look at some pictures from a recent trip we had taken together to Hawaii. One picture showed me standing on the balcony of our hotel room. I not only remembered the balcony overlooking the Pacific, but also instantly called to mind a few details of the evening. One thing I didn't recall was the red macaw perched on the railing just to my right. As I looked at the picture I felt a slight sense of uneasiness. My memory of that evening was quite different from what I was seeing in the picture.

Yes, we had encountered the macaw, but not on our balcony. It had been perched outside the hotel restaurant where we had dined earlier that evening. When I mentioned this fact to my wife she responded: "Don't you remember? After dinner we asked the manager if we could take the bird up to our room so that I could take a few pictures. After taking this picture I brought the bird back down to the restaurant." My uneasiness increased. Sure, I had had a drink or two at dinner, but surely not enough alcohol to result in a complete loss of memory for such an unusual event. Yet how could I doubt what I saw plainly before me in the picture?

When forced to choose between my far-from-perfect memory and the picture, my brain favored the picture, hence my uneasiness. Could this herald the initial salvo of Alzheimer's disease? Then I looked up and caught the shadow of a smile on my wife's face. She then displayed two other pictures in rapid succession.

The first showed a red macaw sitting peacefully on a perch; the second consisted of me standing alone on the balcony. She then showed me how she had cleverly transposed the image of the macaw into the balcony picture, thus creating a convincing picture of an event that had never happened.

As with my response to my wife's picture, marketers and others can alter memories so we become convinced of the reality of something that never happened. Psychologists refer to this as *memory morphing.*

Essentially, memory morphing involves the creation of false memories by various manipulations carried out after the event. Rather than a quirky, inconsequential process, memory morphing is extremely popular with marketers who are now using it to convince customers that they have undergone experiences that actually never took place. The morphing doesn't have to involve pictures; suggestions are often sufficient.

According to research on memory morphing carried out by marketing-oriented psychologists, 25 percent of normal adults will accept the suggestion that they had been lost at age five in a shopping mall and rescued by an elderly person, or that as a child they had spilled a bowl of punch at a wedding. Moreover, these false memories are incorporated into their memories for events that actually did happen at that time in their lives. "With suggestion and imagination a significant minority of people can be led to believe that they had experiences that were manufactured," according to the authors of a paper titled "Make My Memory: How Advertising Can Change Our Memories of the Past."

When I first heard about the shopping mall and wedding experiments I found them hard to believe. But I soon found evidence in study after study that merely encouraging people to imagine an experience increases their confidence that that experience actually occurred. A famous example occurred to no less a personage than the Swiss psychologist Jean Piaget.

The "Kidnapping" of Jean Piaget

Throughout his life Piaget frequently spoke of a vivid memory of an incident from his early childhood. One day while his nanny walked him in a pram down the Champs-Elysées a man leaped out from the bushes and, in an attempt to kidnap Piaget, scuffled with the nanny, who successfully fought him off, but not before he inflicted superficial scratches on her face. Piaget's memory for the frightening event was exquisitely detailed. He recalled the faces of the people at the scene who gathered to express sympathy, the uniform of the policeman, the scratches on his nanny's face, the exact location of the assault. And yet, as Piaget and his family subsequently learned, the episode had never taken place.

Years later, the nanny, after conversion by the Salvation Army, wrote to Piaget's parents and confessed to fabricating the whole incident (including the scratches). Apparently she felt badly enough about her deceit that she returned the gold watch Piaget's parents had given her as a reward for her bravery. Yet even though Piaget now knew that the incident hadn't taken place, he continued to remember it for the rest of his life. Presumably Piaget had assimilated conversations about the event that he had overheard as a child and then stored the pseudo-incident as a true memory. Indeed, the false memory proved so resistant to modification that Piaget ironically termed it "a memory of a memory, but false."

So maybe that shopping mall experience I mentioned a moment ago isn't so incredible after all. If I ask you to imagine being lost at age five, you are twice as likely than controls to express confidence two weeks later that the event actually occurred. You will remember it only too well. *Imagination inflation* is the technical term for this phenomenon. But whatever you call it, the implications are pretty sobering: Marketers can alter our memories in ways that increase the sales of their products while diminishing our confidence in the reliability of our recollections.

An article from *Beverage World* (granted, not exactly your everyday read) entitled "Yesterday, Today and Tomorrow" tells of "many adults," including the vice president of marketing for Stewart's Beverages, who remember from childhood drinking Stewart's root beer from frosty bottles. This memory is false since Stewart's Beverages first began national distribution only ten years ago. Prior to that, the root beer was only available as a fountain drink. What could explain such a discrepancy? Memory morphing. It's likely that the glass bottles, embossed with such phrases as "original," "old-fashioned," and "since 1924," provided customers with the basis for forming a false memory, an illusion of a childhood experience they never had.

In a related experiment researchers were able to manipulate people's memory about a past experience by inducing them to believe that the experience had been more positive than it actually was. Participants drank an unpleasant-tasting orange drink spiked with salt and vinegar. They then looked at advertisements suggesting that the drink was "refreshing." In response, a goodly portion of the participants reported that they, too, had found the drink refreshing.

Other published examples of memory morphing include convincing an adult of having been rescued from drowning as a child by a lifeguard, or having survived a vicious attack by a dog. In one especially fascinating example, a researcher created a fake picture by transposing a childhood snapshot into a photo of his uncle so that it appeared as if the two of them had been sitting together years earlier in a hot-air balloon. Despite the fact that such a trip had never occurred, 50 percent of the adults shown such a fictitious picture recalled a childhood trip in a hot-air balloon. Some of them even supplemented their narrative with vivid details of the trip. When I read of this experiment I thought back to my wife's macaw prank.

Basically, memory morphing takes advantage of the fluidity of our memories, which aren't encoded like videotapes or DVDs that we play back whenever we want to reexperience something from the past. As with Schmolck's findings with people's remembrance of the O. J. Simpson verdict, memory formation and recall involves fluid dynamic processes.

Certainly a moment's reflection on one's own memory confirms that memories are re-created, frequently lost, distorted, or drastically altered (think about those marital spats over the dicey topic of what really happened on a particularly emotionally charged occasion). Further, our state of mind at the time of recall influences our memory. We remember different things when we're feeling down than when we're feeling good about ourselves. Indeed, if we're depressed enough we even alter our memories sufficiently that past events that had seemed happy at the time take on a dark, forbidding tone: "Yes, we enjoyed that week at the beach, but two weeks later the dog died." And, as with the hot-air balloon experiment, people can even be made to believe in the reality of experiences that never occurred.

Consider for a moment some of the implications of these findings about memory. At any given moment, how can you be certain that you're remembering something the way it really happened? At this point in our exploration of memory I hope that I've given you reason to entertain a healthy skepticism in response to someone referring to "the way it really happened" or other phrases that imply consistency, agreement, and objective standards. Such skepticism is certainly justified. The more social neuroscientists delve into the bramble bush of human memory, the less secure we should feel about the reliability of our own memories. And we're not talking Alzheimer's disease here, just normal human memory—which, it's turning out, is more malleable than most of us ever imagined.

Backward Framing

For instance, imagine yourself watching a trailer for an upcoming movie in order to critically evaluate the likelihood of its success. After you've finished your evaluation, you're provided with reviews, both positive and negative, written by professional critics. After that, you're asked about your previously formed judgment of the quality of the movie. Do you think the reviews that you read after making up your own mind about the movie will have any effect on your memory of how you originally felt about the movie? Incidentally, I'm not referring to your suggestibility—your tendency to change your mind about a movie after reading a rave review of it by your favorite critic. All of us are prone to some extent to such authority-driven reappraisals. Rather, I'm referring to pure memory, your recall of how you originally felt about the movie before reading any critical reviews.

Your memory of how you felt about the movie would be changed under such circumstances, according to a study carried out by Kathryn Braun and Gerald Zaltman, both at the time faculty members of the Graduate School of Business Administration, Harvard University. Braun and Zaltman found that people who read positive reviews believed that their initial impression about the movie had been positive even though in many cases it hadn't; readers of negative reviews believed that they, too, had not enjoyed the movie, even though their initial impressions had often been wildly enthusiastic. These findings held true even when the participants were specifically instructed to respond according to their initial opinion, not their opinion after reading the review. What's more, the participants remained unaware that they had altered their memory of their initial impression. In short, the reading of the review had served to transform the participant's memory.

Marketers have come up with a term for how to bring about

such changes of mind as illustrated by shopping mall and movie trailer experiences. They call it *backward framing*, which differs from the usual *forward framing* approach to influencing opinion. In forward framing I tell you ahead of time what you're likely to experience when you see a particular movie. When I do that, you tend to undergo the very experience that I predicted. The same thing happens when you read a positive or negative review of a movie prior to seeing it. Indeed, this positive framing effect is sufficiently robust that many people won't bother to go to a movie if they've read negative reviews about it. (Why do you think Hollywood is so concerned about the effect on sales brought about by negative reviews?)

Backward framing operates in the opposite direction. Here's the definition of Gerald Zaltman, author of *How Customers Think*: "What consumers recall about prior products or shopping experiences will differ from their *actual* experiences if marketers refer to those past experiences in positive ways." In other words, backward framing alters an already established memory: Those participants in the movie experiment didn't seem to remember that their initial impression was positive after they later read a negative review. And even though most of us are sufficiently independent that we can make up our own minds and stick with our opinion despite contrasting opinion from others (even experts), backward framing is sufficiently powerful a technique that marketers are increasingly employing it. (Think back to the orange drink experiment described earlier.)

"Consumers can be influenced to recall prior experience differently and in a manner consistent with a marketing communication, *without being aware that their recall has changed*," according to Kathryn Braun. I've italicized the last portion of that sentence because I find that seemingly throwaway phrase about consumers' lack of awareness of their memory change both fascinating and alarming. We're talking not just about information and facts here, but of emotions and feelings as well. If we can't be certain about how we felt about things in the past, what does that say about our sense of social identity? Our sense of continuity?

I'm reminded of another study based on the O. J. Simpson trial. (Incidentally, no trial in the latter part of the twentieth century has stimulated more research in social neuroscience.) In "Remembering Past Emotions: The Role of Current Appraisals," researchers led by Linda Levine asked the participants how they felt at the time of the verdict. As expected, the responses mirrored those found in the general population: anger, surprise, shock, even glee. But when Levine asked them the same question five years later, the reported feelings were not only very different but also in line with their current feelings. If they were experiencing a general state of anger or resentment, that's what they reported about the verdict. What's more, most of them were unaware that they had described different feelings five years earlier.

Even though Levine's study isn't about backward framing (no attempts were made to influence the participants' responses one way or the other), one could easily imagine designing such a study. You would start by providing "new" information and interpretations of a familiar though controversial event from the past; then you would suggest that all of the "facts" weren't previously available; finally, you would interview figures who were involved in the original events along with "experts" on the event and invite them to express new "interpretations" about what happened in the past.

Such a research project wouldn't differ all that much from what we are exposed to on a daily basis. When we turn on a television it doesn't take long before we encounter somebody with a vested interest in influencing public opinion by using a backward-framing technique (though the person would be unlikely to call it that and may not even be aware of the term). Take, for example, the Oliver Stone film *JFK*, which suggests on little evidence that LBJ played a lead role in the assassination of President Kennedy, or the spate of television shows that appeared several years ago advocating the maverick view that during the last ice age a cataclysmic sea level rise had wiped out all traces of an advanced civilization. This backward framing of the historical record aroused enthusiasm in some circles for learning about and perhaps even discovering this lost civilization hidden beneath the ocean.

In a contemporary example of backward framing Kathryn Braun provided the unsuspecting subjects in her experiment with a rigged Disney advertisement suggesting that children visiting Disneyland would have an opportunity during their visit to shake hands with Bugs Bunny. Actually, such an opportunity would be impossible, since Bugs Bunny is a Warner Bros. rather than Disney character. Nonetheless, about 16 percent of adults who read the ad recalled meeting Bugs during a childhood visit to Disneyland. Among those who hadn't read the ad, no one recalled such a meeting.

"Featuring impossible events in advertising can cause people to believe that they had experienced the events," says Braun. "In some sense, life is a continual memory alteration experiment where memories continually are being shaped by new incoming information. And in marketing the alteration will occur whether or not that was the intent of the marketer. The power of memory alteration is that consumers are not aware they have been influenced. What's more, the marketer can enhance the likelihood consumer memories will be consistent with their advertising messages. Consumers ought to be aware of that power."

Environment-Conditioned Marketing

Given the basic fluidity of memory, neuromarketers aren't having much difficulty convincing some powerful clients (Colgate-Palmolive and Kraft Foods, among others) to fork over large sums ($250,000 is a typical fee) for such things as "environment-conditioned marketing." Minus the jargon, we're talking about marketers controlling the context within which customers learn about a product and thereby positively influencing the customers' feelings about the product.

To take an actual example, a product is freely distributed to vacationers in order to create a mental association in vacationers' minds between the product and positive feelings they experienced during their vacation. If successful, vacationers will remember the product with the same warm feelings they experience when recalling the vacation (assuming the vacation goes well, of course— marketers can control only so much). Memory for the product will thus be "morphed" with the vacation, so when the consumer encounters either the product or an advertisement for it, he or she will associate it with those vacation feelings of relaxation and pleasure.

So what's happening here neurologically? The advertiser is linking separate memory pathways (vacation, free samples of the product) for the purpose of creating a composite memory that will be unique for each potential buyer. Here is Gerald Zaltman describing the process: "A Coke ad depicting teens dancing at a party to a particular style of music activates one neuron cluster, thus producing a particular experience of Coca-Cola. Another ad showing a baby polar bear and baby seal sharing a Coke activates a different neuron cluster, thus producing yet another experience. The two social settings depicted in the ads have different meanings for an individual viewer and thus are likely to activate different internally stored Coke associations."

In Zaltman's book *How Customers Think: Essential Insights into the Mind of the Market* he provides a specific example of how these linkages can be created via some clever misdirection. Here customers are led to consider a particular brand of tire via the creation of an unconscious link between the tread on the tire and safety.

> Marketers should always present a product's functional and emotional benefits closely together in their communications to consumers, even if they do so in highly subtle ways. The idea of extra traction made possible by the design of a tire's tread, when coupled with a picture of an infant, will trigger feelings or thoughts of safety in a more powerful way than showing an adult or simply the benefits of the tread. Some people will notice the tire and the infant at the same time, and the connections (tread and safety) between them are activated from both directions. In other cases, someone may notice the tire or the infant first, but once the tread-to-safety connection is made, further thoughts and feelings about safety will cause more thinking about tread. Thus the design of any communication . . . should be attentive to the partnership between function and emotion. . . . New information becomes more memorable if we "tag" it (that is, associate it) with an emotion.

Bear in mind that Zaltman is not a neuroscientist but a professor of marketing. Yet he has obviously done his homework and understands several important principles of brain operation:

• We understand something new by relating it to something we've known or experienced in the past.

• Memories are usually generalized and based on typical rather than specific examples (I remember I had toast with jelly this morning for breakfast, but I'm less certain whether the jelly was grape or apple).

• Each time we remember something we unconsciously induce subtle changes in the details of that specific recollection.

• Specific techniques can be applied to alter a person's memory in ways that will benefit the marketer.

• An ad will be most effective if it presents information combined with a subtle emotional underpinning.

Brand-Name Loyalists

Can a person's brain be "read" by brain imaging studies so that subsequent predictions can be made about the likelihood that the person will buy a particular product after watching an advertisement? That's the rather extravagant claim that I frequently encountered while researching this book. Basically, it's being suggested in some quarters that a prospective customer be studied in an fMRI machine while he or she watches an ad for a product. On the basis of this scan it's claimed that a marketer could tell whether the person is likely to buy the product. Several pioneering studies are cited to support this view that a "buy button" exists within the brain.

For instance, the brain reacts in a novel way when it's exposed to a famous brand name, according to an experiment carried out at the University of California, Los Angeles. As student volunteers fixed their attention on a computer screen, a mix of words appeared either on the right or the left side, or in either capitals or lowercase letters. The words consisted of either everyday items like *plant* or *car*, brand names such as Sony or Compaq, or nonsense words like *lenkle* and *nevico*. The students had to press a button if they recognized the word as real.

The students readily identified the everyday words when they appeared on the right-hand side of the screen. This finding wasn't surprising, since the left hemisphere of the brain is specialized for reading and preferentially scans the right side of space. With brand names, however, the pattern was reversed: Recognition occurred faster when the brand name appeared on the left side of the screen. This area is read by the right hemisphere, which preferentially processes emotions. Such results are consistent with the possibility that a brand name's power to influence buying decisions results from the emotional responses evoked by the brand name.

As an example of the emotional impact of a brand name, consider the "Pepsi Challenge." In a blind taste test most people prefer Pepsi-Cola to Coca-Cola. But if they're aware of which brand they are drinking, they opt for Coca-Cola. What is the explanation for such a seemingly puzzling finding? To find out, Read Montague, a neuroscientist at Baylor College of Medicine, repeated the challenge with volunteers while monitoring their brain activity. Sure enough, more of the subjects chose Pepsi over Coke. But in addition, Read observed enhanced activity in the ventral putamen, a component of the brain's reward circuitry.

Read then repeated the test, but this time he told the subjects which of the samples were Coke. In this situation, Coke was preferred over Pepsi. Also, a different area of the brain lit up: the medial prefrontal cortex, a key site for executive-style decision making. In essence, the subjects' brains responded differently when they were aware of a brand name.

I came upon an example of brand-name research during a neuroscience meeting in San Diego. C. Pribyl and associates (which included a member of the Gallup Organization) carried out an experiment on customer "engagement," a benchmark for the emotional bond between a current customer and the product. After measuring the fMRI responses of customers of a luxury re-

tailer in Japan, they arranged the responses according to the customer's "emotional attachment to the retailer."

This determination wasn't casually arrived at, by the way. The
investigators used a measure of customer engagement and emotional attachment developed by the Gallup Organization called
the CE11. By this measure, customers could be divided into
"highly engaged" (characterized by a strong emotional attachment
to the retailer), merely "engaged" (moderate emotional attachment), and "disengaged" (little or no emotional attachment).

Customers from the three groups answered simple yes-or-no
statements about the retailer while undergoing fMRI scans of their
brain activity. One finding immediately stood out: The brains of
customers with the strongest levels of attachment and engagement to the retailer were significantly more active while answering the questions about the retailer than the brains of customers
with lower levels of engagement. Rather than taking on an even
distribution throughout the brain, this enhanced activity involved
regions related to thinking, emotional processing, and memory
(the orbitofrontal cortex, temporal pole, fusiform gyrus, and the
amygdala). But were such activations company-specific?

To answer this important question the researchers asked all
participants similar questions about their banks. None of them
showed the same increase in brain activity while thinking about
their bank. Most striking was the strong relationship among brain
activity, attitudes, and spending. The higher the levels of engagement, the more the customer spent with the retailer. "Clearly,
these results suggest that customer loyalty has a profound emotional component, characterized by differential activation of specific emotional centers in the brain," the investigators concluded.

When I first heard of the brand loyalty study I thought, "Why
pay so much attention to the brain functioning of people loyal to
the brand? Why not instead concentrate on what happens in the

brains of new customers?" Actually, as I later found out, marketing experts had already answered my question: The cost of acquiring new customers is about five times the cost of maintaining established ones.

In addition, making brand loyalists out of just 5 percent more customers leads, on average, to an increase in profit per customer of between 25 and 100 percent. In other words, the companies that rely on snagging new customers via megabuck advertising campaigns will always make less money than companies that funnel their efforts into making their established customers happy. (That's what no-questions-asked return policies are all about.) Further, if they do a really good job and establish customer loyalty (engagement), the companies gain a double benefit, because they have snared a lifetime customer whose testaments may entice others to try that brand's product. An important unknown remains, however: How does a company distinguish customers who will become engaged from others who may try the product but fail to establish brand loyalty?

According to research conducted by the Gallup Organization (one of the funders of the fMRI study mentioned above), people stay faithful to brands that engage both their trust and their affection. Moreover, customer engagement is a better way of predicting future buying patterns than customer satisfaction alone. Even better yet, a brain activity "signature" for customer engagement has the potential to create additional lifetime customers—or at least that's the belief held by many marketing futurists.

When asked to produce an informal, reader-friendly version of their technical research, Pribyl and colleagues came up with this:

Imagine that you are the CEO of a major retail company. If you could look into the minds of your customers to see what they are thinking, it would certainly save a lot of guesswork. Savvy marketers now aim higher than merely satisfying customers—they want to create strong emotional bonds with customers instead. Now imagine that the wish of so many CEOs could actually come true. Using a magnet and a computer, you would be able to peek inside the heads of your customers as they think about your company.

At first glance the Pribyl research seems open to the same objection as the Pepsi Challenge: It simply isn't feasible to plunk customers and potential customers into fMRIs for the purpose of registering the emotional appeal of a particular brand or product. But the real point of the study was to evaluate whether a questionnaire designed to address customer engagement and emotional attachment can in fact engage the brain's emotional circuitry. The answer? It does. "What is most striking about these results is that a simple and easily administered battery of attitude items [the CE11] was able to reliably screen individuals whose brain activity differed substantially when thinking about a company or brand."

In other words, the Pribyl results suggest the feasibility of devising pencil-and-paper tests that can identify those who would show strong emotional activation on fMRI. Thus the goal isn't "Let's get all of our customers into fMRIs" but rather "Let's get some of them in fMRIs under controlled conditions, identify a profile, and then use the usual marketing strategies to take maximum advantage of what we've learned." The goal? To create emotional bonds with customers and thereby induce them to buy more.

While all of this appears to be a potentially useful blending of marketing and brain science, the research behind it rests on two questionable assumptions. For one thing, it assumes that our choices arise on the basis of single issues, whereas multiple issues are usually involved.

Suppose I am placed in an fMRI and shown pictures of a new

Ferrari and a new Ford Taurus. I would expect differences would appear in the fMRI findings associated with the two cars. But I think those differences would have little to do with what car I would actually buy. If I splurged and bought the Ferrari, I would impose such financial constraints on my family that the resulting guilt would make it impossible for me to enjoy the exotic car. Thus buying decisions are never (except in the most impulse-driven people) made by single factors but instead are a combination of many factors (the cost of the car, the money conveniently available, the effect on family life, etc.).

Similarly, many areas of the brain are called into play: pleasure circuit components along with executive circuit components, especially those involving the frontal lobes. Indeed, the very organization of the brain itself makes it highly unlikely that anyone is ever going to be able to predict anyone else's purchasing behavior in a specific instance by means of an fMRI. There are simply too many variables. But that hasn't stopped advertisers from using neuroscience to fashion their ads.

Consider a two-page advertisement printed in several national magazines for the Lincoln LS. On the right side of the spread is a large picture of an MRI image of the brain. Above this image is a picture of the interior of the Lincoln. Completing this page are two more brain images along with a picture of an African American woman identified as neurosurgeon Dr. Deborah Hyde. The caption reads: "The brain may be the most complex object in the universe. Deserving of all the myriad stimulation the Lincoln LS can provide."

On the facing page is another shot of the LS. "The Pleasure Neuron: Luxury may be habit-forming and we have the MRIs to prove it" is the headline spanning the width of the page. Below are four short paragraphs of text that contain the following sentences:

After surgery that starts at dawn Dr. Hyde may head out for lunch and shopping. En route, Lincoln LS provides stimulation for all the senses.

Dopamine is released, producing a feeling of well-being. Dr. Hyde adds, "Current research suggests that the nucleus accumbens is very important in pleasure. It's not the cortex, where thought occurs. It's deeper in the brain, where feeling is."

Thus it's possible luxury may be perceived before it reaches the cortex—before you "think" it, you are already enjoying it. And craving more.

. . . the Lincoln LS has been designed to tickle your pleasure neuron.

The message is direct and at the same time subtle. Brain science, it's claimed, has advanced to the point that it's possible to pinpoint exactly what's going on in your brain when you instinctively select one car from among competing models. Simply looking at the car excites a part of your brain called the nucleus accumbens ("deeper in the brain, where feeling is"), which releases a chemical, dopamine, that induces "a feeling of well-being." What's more, the car stimulates your "pleasure neuron." Given all this, how could you *not* buy the car?

The ad is also deceptive on several levels. First, there is no such thing as a "pleasure neuron"; rather, millions of neurons link together to form circuits that become activated during pleasurable states. And even if a pleasure neuron existed, an MRI, which images the brain in millimeter slices, is too crude an instrument to demonstrate it. Second, the mention of the nucleus accumbens adds little to what we already know about what usually goes through our mind when we decide to purchase a car. As a rule, we're drawn to a specific car not because of the specs printed in the brochure or the hype quoted to us by the salesperson but because of the car's looks along with the accompanying mental scenario of how we will appear to others when we're at the controls of this sleek machine. But do we need a neurosurgeon to tell us this?

And do we need to know anything about the nucleus accumbens to realize that sometimes a particular purchase is motivated more by craving than considered judgment ("Can I really afford a new car just now? Probably not, but I really want it and I'm going to make this happen")? In this ad, though, we're encouraged to think of the purchasing process in a new light. The decision isn't really ours to make. In order to justify our purchase (one that in our more reflective moments we may later come to regret) all that we need to know is that our brain's "pleasure neuron" has been stimulated. Indeed, the manufacturers of Lincoln can "prove it."

The Lincoln ad works because it combines a smattering of terms drawn from neuroscience with the respect and even awe that most of us feel toward neurosurgeons. As the ad cleverly assures us, we don't have to feel excluded here simply because we're not neurosurgeons. "It doesn't take a neurosurgeon to appreciate" the advantages of this car, according to a caption for a photo beneath the banner headline telling us that luxury may be "habit-forming."

In this ad we're invited to experience ourselves from the point of view of our brain: We don't choose the car, our brain does it for us. Thus our experience of ourselves as a purchaser is changed in subtle ways. It's okay to let our nucleus accumbens decide for us; in fact, we should allow this because the resulting release of dopamine will produce a "feeling of well-being." Even "before you can 'think' it, you are already enjoying it" and "craving more."

Notice how the ordinarily pejorative term *craving* is given a positive spin. Craving doesn't involve drugs or alcohol or sweets (the usual objects of craving) but something good for us, something we shouldn't resist. By referring to the brain the advertiser seeks to alter the reader's usual assumptions about himself and the cautious, even reflective, attitude he should take toward purchasing a car.

This reference to craving in this ad has some basis in fact, according to the research of Pribyl's team, which is summarized under the title "Neural Basis of Brand Addiction: An fMRI Study." Are you as uncomfortable as I am with the concept that a person's relationship with a brand resembles an addict's relationship to his drug of choice? While that isn't a comparison that gladdens the heart, the findings hint at just such an association: a greater activation of the amygdala among those displaying "the most extreme levels of engagement" with a product. That, of course, is the same response that is observed among addicts: the stronger the emotional bond, the greater the addiction and amygdala response.

As I'll discuss in more detail in the afterword, the suggestion that many of our behaviors conform to an addiction has the potential to exert a profound influence on how we think of ourselves. Am I really "addicted" to Coca-Cola, Delta pens, and Clairefontaine notebooks—to mention just three products I purchase in what others might regard as excessive amounts? Neuroscience is now suggesting that since the same neurochemical (dopamine) is involved in addictions to drugs and "excessive" brand-name purchasing, some degree of emotional bonding, if not necessarily outright addiction, is involved. And marketers are employing several avenues to increase this bonding between their products and ourselves.

Emotion by Design

Music provides an appealing avenue for savvy marketers to create emotional bonds aimed at ensnaring customers. As social commentator Linton Weeks wrote in a *Washington Post* essay on iPods, companies such as Muzak are using the latest findings on the brain "to implant 'earworms'—slang for musical phrases you can't get out of your head." To facilitate this implantation, Muzak consultants sit down with representatives of a company such as Lenscrafters and talk to them about the brand image they're trying to communicate to their customers. They then design a specialized music program that helps "tell" the company's "story."

To get a feeling for these "earworms," go into a favorite store and pay attention to the background music. In most cases the music isn't casually chosen. Every retail store has a logo and a certain look, and Muzak wants to put a musical face on the place, according to Sumter Cox, director of corporate communications at Muzak. That music will vary according to the store's "target" customer.

You're not likely to hear hip-hop while shopping at Saks Fifth Avenue. Nor are you likely to walk through the aisles at Benetton to the accompaniment of a symphony by Brahms. And if the regular customers of these stores did encounter such anomalous musical selections, I suspect the discordance between the store and the music would likely move more of them to the front door than to the checkout register.

Here is Gerald Zaltman, author of *How Customers Think*, on the topic of company "stories":

> To build stories, the brain distributes stimuli (such as product designs, purchase settings, marketing communications,

and other cues) from the thalamus to the cortex and the amygdala. Whereas the cortex is associated with thought, the amygdala (in partnership with the thalamus) is associated with emotion and unconscious processing in the brain. The cortex modulates emotional responses from the amygdala. However, in the case of storytelling, feelings arising in the amygdala also further influence the cortex. Thus emotions and thought stimulate and shape one another.

That's not a bad first take on how understanding what happens in the brain in response to a company's "story" may advance that company's sales. By means of these brain-based neuromarketing techniques, Muzak—along with other firms such as DMX—is creating "audio architecture" (defined by Muzak as "emotion by design") aimed at ensnaring our attention and thereby increasing the likelihood that we'll buy a particular product. And because music can facilitate this manipulative process, we're encountering background music in almost every store we enter.

"This explains," according to Weeks, "why neuromarketers—the lab-coated people who study the brain wave patterns and heart rates of customers under various circumstances—are deeply intrigued by music's effects and how they can be used to manipulate us by the rhythm of a piece, the rise and fall of its structure, certain chord changes."

The marketers are definitely on to something, according to research on the effects of music on the brain. Think back to the last movie you saw that moved you emotionally. If it's out on DVD, watch it again, but this time with the sound track muted during the emotionally arousing scenes. You'll find yourself less emotionally affected. Thomas Baumgartner of the University of Zurich demonstrated this effect in an experiment where volunteers looked at pictures and listened to musical excerpts chosen to arouse happiness, sadness, or fear. The pictures and music were presented to the volunteers either separately, combined correctly (sad picture with sad music), or combined randomly (sad picture with upbeat music).

Baumgartner found that the volunteers experienced the most intense emotion in response to the combined condition, intermediate with pictures alone, and least of all to the music alone. Since we are primarily visual creatures, the primacy of the pictures over sounds should come as no surprise. But music exerted additional influence: The emotional impact of the pictures was greatest when accompanied by music that evoked a similar emotion.

The Good, the Bad, and the Ugly

Moviemakers have long been aware of the emotionally intensifying power of music. Look at a scary movie absent its sound track and you're more likely to be amused than frightened. An appropriate sound track "works" by arousing our emotions. Recently Hollywood has become greatly interested in the use of fMRIs to identify movie sequences that activate the brain's emotional circuitry. The assumption—and it's a big one—is that the lessons learned from a small sample will prove applicable to a large percentage of moviegoers.

An example of this new approach to movies appeared in the prestigious journal *Science* in March 2004. Five volunteers watched the Clint Eastwood movie *The Good, the Bad and the Ugly* while undergoing fMRIs. The investigators found a large degree of correlation between the brain responses of the participants when they watched the same movie segment. One region in the occipitotemporal cortex, the fusiform gyrus, fired up in response to scenes featuring faces, especially full-screen close-ups. Another nearby area, the collateral sulcus, was most active in response to sequences featuring buildings and both indoor and outdoor scenes.

On average, about 30 percent of the cortical activation of one person's brain could be predicted from the fMRI signals from the other person's brain. This high degree of correspondence suggested to the authors that individual brains "tick collectively" and reflect the viewer's degree of attention and the emotional impact of the movie scenes. Consistent with this hypothesis, the correlation between people's brain responses was especially strong during emotionally arousing or surprising segments of the movie, particularly ones containing gunshots and explosions. The authors concluded that "the brains of different individuals show a highly significant tendency to act in unison during free viewing of a complex scene as a movie sequence."

Some movie executives were so impressed by this study that they commissioned a neuroscientist at the California Institute of Technology to measure the brain responses of test subjects while they watched movie trailers. Apparently the interests of the movie moguls and the neuroscientist, Steven Quartz, dovetailed. "We wanted to look at how the brain processes emotions and, since movies induce emotions so powerfully, they were an effective way of doing that," says Quartz.

In a typical experiment Quartz places a volunteer in an fMRI scanner and projects films such as *Casablanca* onto a mirror placed above their eyes. He claims to have found an area, the orbitofrontal cortex at the base of the frontal lobe, that "underlies liking or anticipation. We can look at changes in blood flow in that region to measure how much people are anticipating a movie when they are watching a trailer or how much liking they have. Another region indicates humor. We can look at the whole brain millimeter by millimeter."

Before becoming too excited at the prospect of an fMRI-driven movie industry, however, consider this: While 30 percent of the cortical activation of one person's brain when watching *The Good, the Bad and the Ugly* could be predicted by the fMRI signals from another individual's brain also watching the movie, *the remaining 70 percent could not be predicted.* That large region of nonactivation included most of the prefrontal cortex—the area that is important for critical thought. Thus more than enough cerebral cortex is available for each of us to experience the movie in our own unique way.

There are other reasons to doubt that brain science is somehow going to eliminate the highly personal and inherently subjective nature of movie preferences. One has only to read the sometimes highly discrepant reviews of a particular movie to realize that the laws of science play little role in determining movie preferences.

Gender differences too must be taken into account. For the most part, men prefer action movies, while women like movies that involve relationships. While it would no doubt be interesting to pinpoint areas in the male and female brains responsible for gender-based preferences, such findings aren't likely to be very helpful in changing those preferences. Besides, one doesn't have to be a brain scientist to speculate that movies described by theatergoers as emotionally arousing activate brain circuits concerned with emotional arousal. This kind of circular reasoning is merely a restatement of the obvious in technical jargon.

To discover whether somebody finds a trailer emotionally arousing one could, of course, put him in an fMRI, but why not simply ask him? And if you want confirmation of his response, you don't need to perform an expensive fMRI; you can measure his heart rate, breathing pattern, skin conductance, and other measures of arousal.

Experiments such as the one involving *The Good, the Bad and the Ugly* reinforce the dubious belief that we will ultimately discover within the brain specific areas responsible for some of our most valued but little-understood human qualities (wisdom, love, etc). But brain research isn't ever going to be able to do that. Not only do so such qualities involve many circuits spread throughout wide swaths of the brain, but brains differ from one person to another. Your brain's processing of complex human emotions and experiences isn't going to exactly coincide with how my brain processes them. Think of it this way: In the absence of universally accepted definitions of wisdom, love, and so on, how could we reasonably expect to depict them on a brain image?

9

Neuroeconomics: What Happens in the Brain When We Reason and Negotiate

(and Why Honesty Really Is the Best Policy)

Playing the Ultimatum Game

Would you take $20 if someone offered it to you with no strings attached? Most of us would. But what if your bene- factor revealed that he had earlier received $100 from some- one with the instructions that he should split it with you in any way he chose. As part of this deal, he would have to return the $100 if he could not get you to agree to his division of the money. So, if you agree to take $20, he will pocket $80. How do you feel now?

When I first learned about the Ultimatum Game (the name is based on the fact that the player with the money gives the other player an ultimatum to either accept or refuse the initial offer) I asked many of my friends how they would respond to such a non- negotiable offer. In the process I detected an invariant theme: When the offer was low, outrage and anger emerged based on the sense of being taken advantage of. But from the strictly eco- nomic point of view this doesn't make sense: Your best decision is to maximize your profit by taking the $20. The $80 that the other party gets out of the deal bears no real relevance to your gain of $20.

Indeed, accepting any amount of money—irrespective of the other person's winnings—is, economically speaking, the most rewarding thing to do. Despite the seeming rationality of this economic argument, most people don't respond that way. In most cases, any offer less than 30 percent of the stakes is refused, suggesting that human rationality can't be neatly equated with economic maximization. Instead, most of us place a higher value on fairness and we demand that the money be split more or less evenly. We feel so strongly about gross violations of fairness that most of us are willing to deprive ourselves of a gain in order to punish another person's greed.

Deciding to forgo money in order to punish another person involves not reason, therefore, but emotion. The emotional basis for such decisions becomes especially obvious under circumstances where the participants are known to each other (my wife insists that if she were playing the game with me she wouldn't accept anything other than a 50-50 split; I suspect many other people would demand the same parity when playing with a spouse or partner).

But before concluding that emotion always trumps economic advantage, let's up the ante a bit. Suppose that instead of $20, the offer to you is $20,000 with the other party pocketing $80,000. Would you take that offer, even though you passed on the earlier $20/$80 split? (I know that I would.) No? If not $20,000, would you agree to a $200,000/$800,000 split? Still no? How about a $2,000,000/$8,000,000 divvy? I suspect that at some point, depending on your financial circumstances, the prospect of a financial windfall would offset any sense of unfairness or wish to punish the other person for his greed. At that point, would you be acting rationally or irrationally?

In short, a "rational" decision depends on circumstances.

Raise the ante sufficiently and greed trumps one's sense of fair play ("I value my sense of fair play over a gain of $20 but I'll forget about fair play when the payoff to me reaches $20,000").

The circumstantial nature of human decision making flies in the face of a traditional assumption about economics. According to classical theory, "economic man" will always prefer certain actions over others (e.g., buy low, sell high) once the rationality of these actions has been proven. Thus if given the appropriate knowledge and opportunity, we will always select certain choices and behaviors over others. From this vantage point, economic failures stem principally from ignorance or lack of opportunity (you can't invest what you don't possess; if you are foolish in your investment decisions, you wind up a loser, and so on). You can think of "economic man" (the sexist terminology that still prevails in many sources) as a variation of the "rational man" of jurisprudence: Juries are asked to decide, "What would the rational man do under a given set of circumstances?"

In real life, however, things aren't quite so simple. Not only money but also time is important when it comes to economic decisions. Would you rather have $20 today or $25 in a week? Most of us would take the $20 now. But suppose the decision is between $20 in one year or $25 in one year and a week? According to surveys starting from the 1970s, most people would select the slightly delayed but larger amount. Yet in both instances the delay is the same: one week. Economists have put forth various explanations for this logical discrepancy. "If I'm going to wait a whole year for any payment at all, why not wait another week and get an extra 25 percent on my money?" seems the most persuasive argument.

At this point, let's get very practical and shift our attention from finance to pies. Imagine that we are both hungrily looking at a warm apple pie just taken from the oven. You and I are given the opportunity to take one-quarter of the pie and cut it into two pieces, one for you and one for me. According to traditional wisdom, the safest way to proceed would be for one of us to do the cutting and the other the choosing. Presumably, if I'm cutting the pie I'll do my best to cut it into equal-sized pieces because if I don't, you will select the bigger piece. But suppose I cut the pie so that one piece is conspicuously larger than the other. What might you conclude? From a purely "economic" or even "rational" point of view I have acted against my best interest. My action doesn't make sense. But it only fails to make sense if you insist that an economic model is the only valid one to employ in this situation.

Suppose that I am on a diet or that my doctor told me yesterday that my blood sugar is hovering near the diabetic range. In these instances a small piece (or even better no piece at all) would be the most rational decision that I can make. Or perhaps I'm willing to take a small piece of the pie in the interests of furthering harmony between us. You can no doubt think of other situations where my choice of the smaller piece is the most rational one for reasons having nothing to do with economic factors. Circumstances, not logic, determine rational choice—one of the reasons economic models often fail utterly when it comes to explaining our own and other people's choices.

In addition to circumstances, economic decisions also depend on a particular person's risk tolerance. (Ask any stockbroker.) All of us differ in our willingness to accept risk. Economists have developed some arcane theories to explain risk aversion. Unfortunately, these theories don't take into account some fairly common-sense notions.

For instance, would you choose a guaranteed $600 or a 50 percent chance of winning $2,400? There simply isn't a "rational" decision to be made here that works for everybody. If you're running low on funds, then taking a certain $600 rather than a 50 percent chance of winning $2,400 makes sense; you can put that $600 to immediate use. But if you're financially comfortable, then why not take the risk? If you lose, the $600 loss won't make any real difference in your life, whereas if you win $2,400, you're now in possession of an amount that might really mean something to you.

In other words, risk aversion differs from one person to another depending on each individual's perceptions of what a loss will entail in their own life. In addition—as illustrated by some of the decisions people make during the Ultimatum Game—emotional factors also play a large role in a person's risk aversion.

So wouldn't it be nice to have a brain signature of a person's risk aversion? Neuroscientists are moving closer to achieving this goal. The initial steps toward that goal involved a simple combo of monkeys and fruit juice.

It's All About Me

When monkeys make decisions about, say, the desirability of one fruit juice over another, neurons in the monkey's parietal cortex fire at a rate proportional to the monkey's liking for a particular fruit juice. The more the monkey likes the fruit juice, the greater the firing rate of the parietal neurons. Would a similar correlation between preference and parietal neuronal firing rate hold in humans?

To find out, neuroscientists at New York University placed human volunteers in fMRIs and measured their brain responses while they chose between a sure bet (a 100 percent chance of earning $2) and an alternative bet that varied in both risk and potential payoff (a 50 percent chance of earning $6). This experiment provided two simultaneous measurements: a behavioral measure of a volunteer's risk aversion and an fMRI measurement of blood flow while the volunteers made their decisions. As with the monkeys, activity in the parietal cortex of each volunteer correlated with his or her individual preference.

"Economic theory teaches that individuals have differing levels of risk aversion and that this aversion is reflected in the human brain during decision making," according to Amy Nelson, the chief investigator. "Our findings suggest that the parietal cortex may be the site where individual preferences are represented and may play a key role in decision making. Neuroscience and economic theory have thus converged in our results, revealing a neural correlate of choice preference as predicted by classic economic theory. In answer to the question, 'Are brain areas "risk-averse" in the same way that human behavior is?' we have found evidence that, indeed, they are."

While Nelson has a point, she may be overreaching a bit in her interpretation of her research. Just how much can you conclude about a person's risk aversion from an experiment involving the gain or loss of only a few dollars? A much more interesting experiment would involve thousands of dollars and participants drawn from a broad economic spectrum. Such a spectrum could easily be assembled using volunteers recruited from, say, a casino. But even this unorthodox experiment wouldn't be ideal since, presumably, the most risk-averse personalities never set foot in a casino.

In mini-summary, neuroscientists have discovered so far at least three structures involved in economic decision making: the parietal cortex, the striatum, and the prefrontal cortex. Each connects with the others to form a circuit—there is no single center for economic decisions. Rather, the decisions emanate from the interplay of the various components of the circuit. And at this point, neuroscientists can't be certain that they have identified all of the circuit's components. What's more, additional brain areas may act in opposition to the circuit. For instance, while increased activity in part of the striatum occurs as one person increases his ability to anticipate another person's behavior, an area known as the insula becomes active in people less able to do this. Again, the balance analogy is helpful: The insula on one side of the scale is "all about me," with high activity in this area overpowering the socially savvy striatum on the opposite side of the scale. The end result is an impaired ability to put oneself in another's place. At this point the unspoken question is whether neuroscience can come up with ways of better integrating the cortical and striatal areas, thus making it easier for us to exercise better judgment and more restraint in our economic decisions.

In search of an answer to that question I looked up social neuroscientist Alan Sanfey and asked him about what takes place in the brains of people playing the Ultimatum Game. Dressed in a sport shirt, chinos, and black Buddy Holly–style glasses, Sanfey might strike you as either nerdy or cool, depending on your interpretive inclination at the moment. He says, "The Ultimatum Game involves a conflict between a person's emotional and cognitive responses. The emotional response arises principally from the anterior insula, an area that is responsible for the negative visceral ('gut') responses characteristic of disgust. Offers that are rejected in the Ultimatum Game are accompanied by activation of the anterior insula. Thus moral indignation may be a form of disgust. In fact, anterior insula activity increases in direct proportion to the unfairness of the offer."

Brain scientists have known for years that when people experience disgust their anterior insula fires up. For instance, if you smell rotten eggs or rancid butter, activity increases in your anterior insula, according to the results of an fMRI study carried out by Christian Keysers, a neuroscientist at the University of Groningen in the Netherlands. The same thing happens if you watch a video in which an actor sniffs at a glass and conveys by his facial expression that he's just whiffed something disgusting. So Sanfey's finding that anterior insula activation occurs whenever we experience moral indignation is consistent with common experience. When we express moral indignation we often express ourselves in phrases like "His lecherous behavior was truly disgusting."

The anterior insula isn't the only area activated in the Utimatum Game. The dorsolateral prefrontal cortex (DLPC), associated with goal setting and keeping to goals, is also activated. According to Sanfey, "The DLPC attempts to overcome the anterior insula and says, 'Let's try to make some money here.' Think of the anterior insula and the DLPC as in conflict with each other. The DLPC wants to come away with some money rather than nothing, while the anterior insula wants to reject offers that seem unfair. What's needed here is a mediator. That mediator is the anterior cingulate, which balances the disgust and rejection of the anterior insula with the understandable 'rational' desire of the DLPC to make some money."

But this internal negotiation doesn't always work. As mentioned earlier, players in most parts of the world where the game has been played will reject any offer that is less than a third of the total amount of money. Interestingly, this doesn't happen if the player is knowingly "negotiating" with a computer. "People don't get angry and disgusted with a computer the way they do with another person and are therefore more likely to take whatever they can get," Sanfey told me.

During his discussion Sanfey mentioned another game, the Dictator Game, which is similar to the Ultimatum Game but with one essential difference. In the Dictator Game the second player doesn't have a choice to accept or reject the offer but must take whatever is offered, no matter how seemingly unfair the offer.

In the Dictator Game there is no decision to make and therefore no conflict. As a result, neither the DLPC nor the ACC is activated. "Here we have an indication that while even very simple decisions activate emotional center of the brain, the DLPC and the ACC aren't called into play when there is no real decision to be made," Sanfey explains. "Even though our economists and political leaders don't seem to be aware of it, every decision—if it's a real decision and not something foreordained like in the Dictator

Game—involves conflict between parts of our brain involved in both thinking and feeling." Smiling wryly, Sanfey mentions that the result of this conflict isn't always easy to predict. "In the Ultimatum Game women often offer attractive men more than 50 percent of the money."

In summary, offers in the Ultimatum Game activate specific brain circuits: the anterior insula (AI, a center for disgust, both physical and emotional); the dorsolateral prefrontal cortex (DLPC), associated with setting and keeping to goals; and the anterior cingulate cortex (ACC), involved in resolving situations of mental conflict. Activation of the components of this circuitry accounts for the inner feelings aroused in the second player in response to what he considers a paltry offer. Thus the circuit involves disgust from the anterior insula, an attempt at appeasement and acceptance of the offer by the DLPC, and finally a decision on the part of the ACC. The ACC may decide to accept the offer or side with the anterior insula. If the latter, the offer is rejected with disgust and—thanks to the spread of the impulse to the other components of the limbic system—responded to with anger or annoyance.

Other fMRI studies have concentrated on people choosing between short-term and long-term gains. Differences in brain activation occurred depending on whether the person demanded an immediate payoff or was willing to exhibit some patience and get the money at some time in the future. Any guesses about the fMRI patterns? It should be no surprise to learn that activation of parts of the limbic system occurred when the person chose the immediately available reward. In contrast, when the person chose to wait for the money, activation occurred in regions of the prefrontal and parietal cortexes. Further, the longer the person was willing to wait for the payoff, the greater the prefrontal-parietal activity.

The most interesting experiment on economic decision making involved a modification of the Ultimatum Game. In this version, the proposal is made to two people rather than just a single responder. If either of the two responders accepts the offer, the money is divided between the proposer and the responder who accepts the offer, or split three ways if both responders accept the offer. As with the original game, nobody gets any money if neither responder accepts. Brooks King-Casas and his team from Baylor that carried out the experiment discovered that the addition of the second responder typically leads to lower offers being accepted—a clear benefit for the proposer. That's because in the three-person version of the game each of the responders thinks of herself as competing with the other ("If I don't accept that low offer, the other person may take it and I'll end up with nothing"). If you've ever been involved in a bidding war for a house or offered more than you should have at an auction, you know the internal state that accompanies competitive purchasing. In the three-person Ultimatum Game similar feelings are invoked, but here the idea is to get something, no matter how small, rather than end up with nothing. Notice how moral-ethical sentiments ("I won't take anything less than a fair offer")—the dynamic that ruled the one-responder Ultimatum Game—yield in the two-responder version of the game to narrow self-interest.

The Baylor researchers also found in the two-responder version an accompanying change in brain activation: decreased activity in the medial prefrontal cortex (MPFC), an area associated with moral decision making. Change the situation and you change the rules of economic exchange. One-on-one interactions lead to an emphasis on fairness; the presence of an additional competitor shifts the emphasis toward one-upmanship. And with this shift comes activation in a different part of the brain.

So what should we make of the fact that several different brain areas seem to be involved in economic decisions, depending on the circumstances and interactions of the people involved? Remember that the brain is composed of interrelated circuits, with many different structures constituting these circuits. At the moment we don't know all of the circuits, nor can we identify all the brain structures comprising these circuits. We're talking here of a science that is still in its early stages. But even at this early stage we know that different circuits, many of them involving the same brain structures, provide the foundation for our decision making.

The Cynic, the Trustee, and the Banker

Economic decisions are made not in laboratories, of course, but during everyday situations involving people not only interacting with other people, but also *trusting* them. Indeed, trust underlies economic transactions on every level, from the individual purchase to global economic exchanges. One nation must rely on the word of another nation to provide raw materials or purchase those materials at an agreed-upon price. When you're buying a product you trust that it will live up to advertised claims. You're also sizing up the salesperson for a sense of trustworthiness that will help you make your purchasing decision.

There is an additional reason why it pays to cultivate trust. According to recent brain research, distrust is dangerous because it increases the likelihood that the distrustful person or nation might take aggressive action. An aggressive response toward the distrusted party is especially likely when men are involved. (A reason to favor women diplomats, perhaps?)

The experiment demonstrating this gender difference involved two hundred men and women paid $10 apiece to allow the experimenters to draw four tubes of blood from their arms and then to participate in a computer exercise designed to measure their response to another person's distrust. After taking seats in a large computer lab, the participants were matched up in randomly chosen pairs. The players did not know the identity of their partner. One-half of the players (the investors) then had the chance to send by computer none, some, or all of their $10 to their partner (the trustee). Whatever amount they sent was tripled: If they sent $4, the trustee wound up with $22 ($4 tripled plus the original $10). At this point, the trustee in the pair learned the amount sent to him or her by the investor. The trustee then had to decide whether to send some amount of the winnings back to the investor or keep everything.

A cynic would expect either that the investor wouldn't send anything (the safest bet if you don't trust the other person to send you at least some portion of the earning) or that the trustee would make off with whatever amount the investor is foolish enough to forward. Obviously, the only reason the investor would send money in the first place is that he trusts his unknown trustee to return some of the winnings. If he doubted that his trustee would do this, he wouldn't send anything. And since the transactions take place anonymously, interpersonal factors ("He doesn't look trustworthy to me") play no part. Thus in this clever experiment distrust is measured by the amount of money the trustee does *not* receive.

Well, the cynics turn out to be wrong. Eighty-five percent of the investors forwarded some of their money to their partners, and 90 percent of the trustees sent some money back.

Next came the second leg of the experiment: determining the biological basis of trust and whether it differs between men and women. After each decision maker reached his or her decision about what if any money to send, the researchers took a second sample of blood and compared it with the first sample in regard to the levels of two hormones that circulate in the blood of both men and women.

First a few words about the hormones. Dihydrotesterone (DHT) is a bioactive breakdown product of testosterone with many physical effects and associated with primitive emotional responses such as explosive anger and physical aggression. Oxytocin, in contrast, is a "feel-good" hormone that is relaxing and de-stressing. It is made in the hypothalamus, a small but critical collection of nerve cells located deep in the brain that controls biological reactions including body temperature, hunger, and thirst. For many years oxytocin was considered nothing more than a reproductive hormone that in women stimulates uterine contractions during labor and later induces milk secretion. But it also serves an additional sexual function: In both women and men oxytocin is released during sex. But oxytocin isn't just about sex. It also plays an important role in attachment, motivating mothers for the demanding task of child rearing. In addition, oxytocin is a powerful stimulus for enhancing socialization. Animals injected with oxytocin into their amygdalae tend to cluster together and touch each other more often.

Returning to the trust-distrust experiment, the researchers found different responses among men and women. Men trustees showed a strong aggressive response when they perceived signals of distrust, and their levels of DHT increased between the two blood samples in direct proportion to how much they were distrusted (measured by the little or no money sent to them from the investor). Among women the level of DHT did not rise in response to distrust. Oxytocin, however, increased in both men and women trustees in response to trust (receiving money).

If oxytocin is a kind of "trust hormone," then what would be the effect of giving someone a dose of oxytocin? Would that have any effect on that person's level of trust? That very question formed the basis of an experiment involving 178 students drawn from universities in Zurich. Each participant received either a placebo or a synthetic version of oxytocin delivered by nasal spray. They then played the Trust Game described above.

Of twenty-nine volunteer recipients of oxytocin, 45 percent invested the maximum amount and thus showed maximal trust. So trusting was the oxytocin group that only 21 percent invested less than three-quarters of the amount available to them. In contrast, among those not receiving oxytocin, only 21 percent invested the maximum, while 45 percent invested at low levels. Overall, the investors who received oxytocin invested 17 percent more than investors who received the placebo.

While the Zurich findings support the trust-enhancing theory about oxytocin, it left unresolved a very important question: Does oxytocin promote social interaction and trust or does it simply increase risk taking? To answer this question, the senior researcher, Ernst Fehr of the University of Zurich, repeated the experiment but with the trustees replaced by a computer program. Under these conditions, the oxytocin and placebo groups performed similarly—not so surprising, really, since it's difficult to socially interact with a computer program and even harder to develop a sense of trust.

Such experimental results suggest intriguing possibilities. Could an unscrupulous con artist increase another person's trust by surreptitiously administering a dose of oxytocin ("You sound a bit congested. So before we talk business, try this new nasal decongestant that will get you breathing better in seconds")? Based on the experiments described above, the chemical induction of misguided and inappropriate trust could actually take place.

And what about people who are naturally distrustful of others? Are they that way because of low levels of oxytocin? Conversely, consider a person who is repeatedly bilked because of his willingness to trust people he doesn't really know. Does he have an excess of oxytocin in his brain? At this point, answers to these questions aren't available. But some practical consequences are already apparent.

Think of the experiments on trust as demonstrating the existence of two physiologic "levers" in our brains (oxytocin and DHT) that activate in response to our interactions with other people. By creating an atmosphere of trust we enhance the oxytocin levels in the brains of those we come into contact with, and vice versa. Alternatively, if we signal distrustfulness, we activate that second "lever" and increase DHT (at least among men) along with the accompanying likelihood of an aggressive response directed toward us.

Left at this, the best response would seem to involve expressing maximal trust toward others. The problem, of course, is that real life isn't like those experiments involving anonymous encounters during which we're unable to evaluate another person's trustworthiness. Some people don't seem very trustworthy, based on their previous behavior. Withholding our own trust would seem advisable when dealing with such a person. But we should always be cautious about acting on our perception of untrustworthiness because we now know, on the basis of experiments such as the one just described, that we're inviting a retaliatory aggressive response. So what course of action would you suggest in this catch-22 situation?

For my part, I've decided to hedge my bet. I plan on trusting other people, even complete strangers, but only up to the limits of what I'm prepared to lose. I'm betting on that rise in oxytocin in the other person's brain in response to my trusting response. And if I'm dealing with someone who abuses my trust, then things will still turn out, since I'm limiting my potential losses. Granted, this isn't a perfect solution, but it does enhance those brain mechanisms we possess that foster cooperation and limit isolation. If I act in a trustworthier manner and thereby elevate that oxytocin level in the other person's brain, he or she is more likely to trust me, thus bringing about the same beneficial effect in my brain.

Revenge Is Sweet

In case you're still not convinced that trustful reciprocation is the best modus operandi, consider this variation on the experiment just described. Player A and Player B both have the same amount of real money ($100, for example). Player A has a choice of either keeping her money or giving it to Player B. If she selects the latter choice, the amount given will be quadrupled, leaving Player B with $500 (his own $100 plus the $400 from Player A). At this point in the game Player B is asked to decide whether to give back nothing or half of his bounty, thus earning each player $150.

In almost all occasions in this experiment, conducted by five neuroscientists including Ernst Fehr and an economist, Player A gave away her money, trusting that Player B would share the profits. But sometimes Player B proved untrustworthy and did not share. Player A was then given a minute to come up with a response.

Several options were available to Player A. In one, the "intentional and costly" (IC) option, punishing Player B came at a monetary cost to Player A. In the "intentional and free" (IF) option, punishing B didn't involve any cost to A. While Player A mulled over her response to this violation of trust, the researchers monitored her response with a PET scan. In both the IC and IF options the PET scan revealed increased activity in the caudate nucleus, an area beneath the cerebral hemispheres normally involved in processing rewards. But even more interesting was a second correlation: the greater the activation of the caudate, the more likely it was that Player A would be willing to punish Player B even when the punishment came at a personal cost.

"A higher caudate activation reflects the greater expected satisfaction from punishment, which, in turn, causes higher investments in punishment," write the authors of the study. "Many people voluntarily incur costs to punish violations of social norms.

Our findings indicate that people derive satisfaction from punishing social norm violations and that the activation in the caudate reflects the anticipated satisfaction from punishing defectors." Put more informally, revenge is sweet.

The Zurich study helps explain that frisson of pleasure we all experience when first learning that corrupt corporate officers have been sentenced to prison for lying and cheating (the intentional and free option, IF). It also sheds light on seemingly inexplicable behaviors such as road rage (the intentional and costly option, IC). Driver A may, foolishly, risk his own life in a high-speed confrontation with Driver B if Driver A perceives some form of transgression, such as if Driver B cuts off Driver A, who is then forced to brake suddenly in order to avoid an accident. Passion and emotion, not reason and logic, are in play in such situations. Even though Driver A knows full well that retaliatory behavior is unnecessary, dangerous, and potentially costly, he'll still seek revenge—even at personal cost—for a real or imagined violation of the rules of the road.

So keep the Zurich study in mind if you're ever tempted to act in a way that another person may perceive as unfair. Because of that caudate activation, people can't always be counted upon to act reasonably when they feel violated or otherwise taken advantage of. In the words of one commentator on the study, Brian Knutson, "instead of cold calculated reason, it is passion that may plant the seeds of revenge."

10

The Perils of the Neurosociety

The "Office-Obsessed" Employee

Judging from developments over the past decade, cultural and social changes will continue to bring about alterations in the brain. In turn, the resulting brain alterations will alter our culture and our society.

For example, consider this advice written by the president of a company dedicated to job search counseling: "Do you really want to enjoy your summer vacation? Then keep your cell phone handy and stay on top of those emails." As the counselor explains, "In a fiercely competitive global economy, where down sizing is a fact of life, no one can afford to be completely out of touch for an extended period." His advice? Rather than forgetting about work while on vacation, he suggests becoming an "office-obsessed" employee. "An office-obsessed employee is recognized by the employer as someone who puts the needs of the company first and therefore will likely survive any workforce reductions. 'Wish you were here' should not just be something you write on a post card; it is what your co-workers and superiors should be thinking when you're away. How important do you want to be at the office—not missed or missed a lot while on vacation?" Accompanying this advice is a picture of a woman sitting on a beach chair and typing on a laptop while the ocean waves lap at her feet.

You don't have to be a psychologist or psychiatrist to recognize the basic lunacy of the above advice, at least according to traditional ideas of what a vacation is supposed to be all about: an opportunity for relaxation, letting down one's defenses, moments of reflection, perhaps even achieving insights such as "It's time to look for a job with a more understanding supervisor." But I think the real damage wrought by such a redefinition of a vacation affects the brain. And I'm not talking about the tension and inability to relax that must inevitably result from remaining "office-obsessed" while on vacation, nor the insecurity and anxiety that such advice can generate in people concerned about surviving "workforce reduction." Although such factors are important, I'm referring to the changes in our brain functioning brought about by technology.

Take that woman sitting on her beach chair and tap-tapping on her word processor while shutting out the sights around her in order to focus on office matters. An fMRI of her brain would show activity in all the wrong places—not the visual area, involved in watching her six-year-old playing in the sand, nor the auditory area, stimulated by the sound of the waves, but in the prefrontal area as she runs over in her imagination various office scenarios. Communication technology is bringing about a profound transformation in this woman's brain: She can no longer compartmentalize her life in the traditional manner, distinguishing between work and leisure and between public and private space. If she buys into the advice to become an "office-obsessed employee," she will eventually become convinced that her basic worth and identity derive solely from her job. And if she loses that job, the experience may prove devastating.

Of course, "office-obsessed employees" have always existed. What's new is how technology is facilitating such behavioral aberrations (and that's what they are). "Technology has become so portable and so affordable that there really is no excuse anymore for not staying in touch. We live in a fiercely competitive world where employers are not going to be very understanding if a piece of business is lost because you are purposely unreachable," says the job search counselor. Given the increasing prevalence of this mind-set, is it any surprise that the number of people in the United States who claim they are overworked has risen from 28 percent in 2001 to 44 percent in 2004?

Other changes in the workplace are also altering the ways our brains normally function. Take, for example, the current efforts in some businesses to put a human face on social communication without the expense and problems that may arise when working with real people. Such efforts recruit different brain circuits than those used during normal person-to-person social interaction. For instance, "communication" with an airport check-in machine is a very different experience from standing at the counter and exchanging pleasantries with a ticket agent. It also involves different brain pathways.

Numerous other examples readily come to mind. Dial 411 for telephone number information and you will probably end up answering a series of automated questions; call your doctor's office and the recorded voice will tell you to call 911 if you think your situation is an emergency and, if not, to "consider the following options by pressing the appropriate key on your phone."

Thanks to an emphasis on speed, efficiency, and economy aimed at eliminating "unneeded" employees, face-to-face communication in our present business-oriented culture is becoming rare. This alters how our brains process empathy and other emotional responses. And things can be expected to get worse. For instance, how do we emotionally respond to a robot, especially one that *appears* to be emotionally responding to us? This is not an idle question since robots are presently under development that will be capable of producing humanlike emotional expressions—smiling, frowning, and so on—and react to the facial expressions of people. In the planning stages are robotic toys that will track the eye gaze of the human player and change their response according to the player's focus of attention. And Hollywood has progressed from an emphasis on special effects to the creation of the most powerful special effect of all: a computer-generated megastar that is indistinguishable from a real person.

Defining one's personal reality will become increasingly difficult thanks to ever-evolving advancements in communication technology. Thanks to the 24/7 worldwide coverage provided by CNN and other news outlets, we are sometimes more actively involved in events happening at far distant locations throughout the world than with what's happening in our own neighborhood. As a result, televised coverage of an election in Iraq or a bombing in London may exert an unwarranted influence on our thinking and our emotions.

Typical of this response is one of my multiple sclerosis patients who suffered a relapse during Hurricane Katrina. Although she lives far from the areas most affected by the storm and doesn't personally know anyone living there, she found herself drawn more and more into the events happening in New Orleans. "Since I work at home I had to keep fighting off the urge to turn on the television and watch what was happening," she told me. As she watched the grim scenes unfolding on CNN, she began feeling increasingly anxious and depressed. When specific symptoms of multiple sclerosis began to emerge (numbness, deteriorating coordination) she turned off the television and refocused herself on her work. Over the next several days the MS symptoms remitted without specific treatment.

While my patient represents an extreme response, none of us is totally free from the influence brought about on our brain's functioning by communication technology. Increasingly, images of destruction and human suffering occurring sometimes halfway across the globe are activating emotional circuits in our brain that normally are activated by events happening right here. I'm not suggesting we ignore world events. But it is advisable to put them in perspective. While televised images do not involve the same kind of reality as personal experience, both televised images and personal experience exert a powerful influence on our brain's emotional circuitry.

Eventually, repeated exposure to emotionally arousing images depicting happenings that one can personally do little about induces emotional burnout. "Donor fatigue" is the term used by experts on disaster relief. When the number of natural disasters that are intensively reported on increases (as happened with hurricanes, earthquakes, and a tsunami between December 2004 and October 2005) people's compassion and attention start to wane, as does their willingness to provide money for relief efforts.

As the overexposure to catastrophes escalates further, the ensuing burnout comes accompanied with anxiety and depression. Like it or not, there really are limits on our capacity for empathy and compassion. This capacity will be further tested as we approach the time in the neurosociety when anything happening anywhere on the globe will be only minutes away from depiction on our television screens.

Identity by fMRI

Take a moment to reread the description of Mario Beauregard's research on pornography discussed on page 125. Now imagine yourself forced as a requirement for employment to have your brain activity monitored by fMRI as you watch various erotic film clips.

Even though your prefrontal cortex may remain in charge, successfully counteracting the arousal patterns originating in the pleasure circuits, those pleasure circuits will nonetheless announce themselves via activation patterns that can be easily picked up by an fMRI. Moreover, those patterns may contain some surprises.

I'm thinking now of a man I once saw as a patient who on rare occasions during his life experienced erotic homosexual dreams. Usually he just shrugged off these atypical experiences and went on with his life. He considered himself totally heterosexual, had never engaged in any homosexual acts, and had been successfully married for over two decades. What would happen if that man took the fMRI test while looking at erotic film clips? I believe it's possible—indeed, likely—that activity might appear in the emotional pleasure circuitry during the homoerotic scenes. Does that identify him as a latent homosexual? Or, instead, should one take a more behavior-oriented attitude toward his sexual identity, in which only his behavior matters?

While most people would favor a more behavioral approach, the man's employer may consider the fMRI results as more confirmatory of his sexual orientation than either his personal beliefs or his lifelong history of strictly heterosexual behavior. Perhaps the man himself may reach the same conclusion, entering into a midlife crisis about his sexual identity. So what should be relied upon to determine the man's "real" sexual orientation—avowed sexual affiliation and lifestyle, or the revelations of an fMRI scan?

In short, the answers provided by neuroscience aren't necessarily definitive. Indeed, they often aren't as valid as information gathered from traditional sources (asking questions, making behavioral observations, etc.). Keeping that fact in mind won't be easy, however, especially when some are claiming that neuroscience will reveal definitive and reliable information about people.

As I have emphasized throughout this book, brain patterns for complex and sophisticated levels of thought and behavior involve multiple circuits, many of which vary from one person to another. That's why it's so risky and often just plain wrongheaded to attempt too rigid a localization of complex and multidetermined behaviors to specific locations within the brain. Elizabeth Phelps, whose work on the brain and racial prejudice I discussed on page 89, even goes so far as to suggest that the scans do not reveal anything about people's racial attitudes that doesn't show up in behavioral tests. Indeed, the tests are routinely used to work out what the brain scan images actually mean!

Even worse are the ongoing attempts to predict behavior on the basis of fMRI or other brain imaging devices. Who might turn out to be violent? Who might be a pedophile? Who has the makings of a stalker? Law enforcement agencies and insurance companies will be especially interested in any technology with the perceived ability to supply answers to such questions. And even though many of the neuroscientists are themselves urging caution ("There is nothing absolute in our data," as one of them notes), this may not stop insurance companies from denying coverage if an applicant refuses to undergo an fMRI aimed at uncovering potential future health problems, or stop employers from insisting that job applicants undergo scanning in order to assess their suitability for a job. (At the moment the costs of such dubious ventures would be prohibitive, but as the technology advances, that obstacle will soon be overcome).

Part of the appeal of fMRIs and other scans is the seductive nature of these multimillion-dollar machines. "The temptation to use brain scans where no other data exists is likely to increase, even though the information contained within scans is often imperfect," as the prestigious science journal *New Scientist* concluded in an editorial in July 2004.

The use of imperfect information about the brain to draw unwarranted and extravagant claims about personality and behavior has a colorful history that we should be wary of repeating.

Neophrenology

Nineteenth-century phrenologists believed that the brain was compartmentalized, with every function potentially localizable to a specific site. This belief stimulated efforts to identify the brain areas responsible for falling in love, laughing at a joke, or tasting a glass of scotch. To the phrenologists, such personality and character traits could be "read" by palpating or measuring the shape of the skull. Spearheading this effort was an Austrian physician Franz Joseph Gall.

Today's neophrenologists have moved further inward with the aid of various imaging devices to concentrate on activity occurring within the brain itself. But the aim of these very different approaches is the same: to identify "centers" within the brain that are responsible for character and personality. Despite the gap of almost two centuries separating these two approaches, striking similarities exist between them.

Gall and his followers argued that the brain was composed of faculties that after repeated use would bulk up (hypertrophy); if neglected, the faculties would shrink. Thus the brain could usefully be compared to a system of muscles. Just as muscular development varies from one person to another depending on experiences, habits, interests, and so on, the brain's size and general configuration would also vary since each person possesses a unique pattern of faculties. Further, the phrenologists claimed that the skull overlying the brain would bulge out to accommodate the enlarged parts of the brain. In order to read a particular brain, therefore, it was only necessary to measure the bumps on the skull and thereby discover which parts are enlarged and therefore, by implication, the dominant faculties.

Today we recognize the error underlying the central tenet of phrenology. While it's true that on a microscopic level the number of nerve cell connections varies with use, this doesn't lead to increases in brain tissue that can be measured with anything as crude as calipers or the automated phrenology machines that were the rage in the nineteenth century. Besides, the overlying skull is firm and unyielding and, except during early infancy, doesn't change in configuration based on any localized enlargements.

The phrenologists were correct about one thing, however: It's possible to localize certain functions to specific brain areas. But what functions? On this critical point, the phrenologists overreached.

At the moment I'm looking at a phrenological chart dating from 1842. Drawn on the surface of the skull are numbers ranging from 1 to 35 with the numbers corresponding to parts of the brain responsible for "personality organs." Included are self-esteem, hope, wit, amativeness, and combativeness. For the most part, 164 years of subsequent research on the brain have failed to discover many correlations between such personality traits and brain structure. This hasn't stopped neophrenologists from making such claims.

Remember the enthusiasm during the 1960s and 1970s for dividing everyone into right-brain and left-brain types? The left hemisphere was supposedly analytic (reading, writing, arithmetic) and the right hemisphere synthetic (responsible for intuitive and creative skills). This evolved from research on the split-brain patients, mentioned in Chapter 2, whose hemispheres were surgically divided. Inspired by this research, popularizers suggested that we learn to draw on the right side of our brain and trust our right-brain intuitions when evaluating people. We don't hear much along these lines anymore, and for good reason.

"Research with normal subjects shows that individuals vary greatly in the development of asymmetrical functions of their brains and in the ways the hemispheres are habitually activated," notes Colwyn Trevarthen, who did much of the early right-brain/left-brain research. "Humankind is not only very diverse in outside appearance: minds in any race, even in a single family, are inherently different in cognitive bent."

The right-brain/left-brain distinction was "the dying kick of phrenology," according to Richard Gregory, editor of *The Oxford Companion to the Mind*. Well, it's not quite dead yet. The development of PET and fMRI has led to the development of neo-phrenology. Instead of linking thoughts and behaviors with specific locations within the brain, the neophrenologists offer us a promissory note: "Not now, but later—very soon. In fact, let me tell you about this research we're now carrying out that might make it happen tomorrow." Fortunately, not everyone is making such extravagant claims.

"The idea that we can accurately identify specific areas of the brain as responding to a particular stimulus is very alluring, but is reminiscent of phrenology," says marketing professor Gerald Zaltman of Harvard University. "We cannot 'read' specific thoughts; we can only tell that brain areas known to be associated with particular kinds of thought and feelings are being activated. Inferences can be made about the quality of those thoughts by knowing what other areas are coactive or active before and after another area, but this still does not identify the exact thought or feeling."

This doesn't mean that we can't learn a lot about human emotions and thinking from fMRI scans. We all share a common brain—though this is differently organized in each of us based on our genetic inheritance, our experiences, and no doubt other causes yet to be discovered. Thus we shouldn't be surprised that our emotional circuitry is much the same from one person to another. This is especially true when it comes to experiences and traits that enhance survival, such as feeling appropriate fear. While we've learned over the last decade that fear is linked with specific brain sites, other emotions such as joy and happiness aren't at all localizable. That's because joy and happiness involve millions of neurons in widely scattered areas throughout the brain. Thus there isn't a joy or a happiness "center" that can be shown to light up by an fMRI.

A modern map of the brain correlating function with location is far more prosaic than the phrenologist's map. Instead of personality characteristics, it depicts centers for vision, hearing, touch, movement, sensation, and the speaking and understanding of spoken and written language. Generally, the more focused the inquiry about human thought and behavior, the more accurate the results. Thus neuroscientists have discovered multiple areas in the brain involved in speech. A less focused inquiry such as "Where is the center within the brain for sexual attraction?" produces less satisfactory results.

Researchers on human sexuality are currently carrying out fMRI tests aimed at discovering what happens in the brain during orgasm. Not surprisingly, they're finding that much of it involves components of the brain's emotional circuitry: the nucleus accumbens, the amygdala, and the hypothalamus. Two other findings stand out. First, the brains of men and women look much the same during sexual excitement. Second, multiple areas rather than a single "sex center" are involved. But since one area, the hypothalamus, is especially active when volunteers think about their sex partners, some researchers are suggesting a one-to-one correlation.

"Love is just a biological mechanism," asserts Andreas Bartels, a brain researcher at the Max Planck Institute for Biological Cybernetics in Tübingen, Germany. And since the hypothalamus is also a principal area in the brain where the sex hormone oxytocin is absorbed, Bartels believes we may only be a few years away from a reliable means of chemically creating sexual attraction: "You can inject an animal with oxytocin and make it pair-bond with a stranger. All we need to do is apply it to humans." In this lesson in applied neophrenology, love, sex, and orgasm can be reduced to a chemically fired-up hypothalamus.

As another example of the oversimplification espoused by the neophrenologists, consider how you would interpret the many studies establishing a link between fear and the amygdala. Although the studies are convincing (the amygdala really does light up when we're afraid), a problem remains: What does that activation of the amygdala in fear-inducing situations actually represent? At the most basic level, it indicates the appreciation of danger. But to stop at this is to leave the really interesting questions unanswered. What should one do in response to the fear? Is it wiser to stand and fight, or is it best to hightail it out of there? Activation of the amygdala doesn't tell us. That's because the decision to fight or flee isn't a reflex response resulting from the activation of a "fear center" or even a "fear circuit." If that were true, everyone's response to a specific threat would be the same. But people vary greatly in their responses to identical situations. Some flee, while others stay and fight—a decision based on the frontal lobe's evaluation. In itself, a highlighted amygdala on an fMRI isn't much help if you want to understand what fear is all about.

"A Whole Bunch of Psychobabble"

Politics is currently providing another, seemingly unlikely venue for neophrenology. During the 2004 election campaign neuroscientists at the UCLA Neuropsychiatric Institute performed fMRI scans on the brains of ten Republicans and ten Democrats while they looked at political advertisements and images of the presidential candidates. When the participants in the study were viewing their preferred candidate, a portion of the prefrontal cortex (the ventromedial area at the midline and toward the base) was activated. Looking at the opposing candidate, in contrast, activated another prefrontal area toward the top and outer part (dorsolateral), along with the anterior cingulate.

After the election Joshua Freedman, a psychiatrist at UCLA, authored an op-ed piece for the *New York Times* describing the research. Activation in the ventromedial prefrontal cortex indicated "strong instinctive feelings of emotional connection," he said, while activation of the anterior cingulate and dorsolateral prefrontal cortex suggested "cognitive and emotional conflict" and an attempt to "suppress or shape emotional reactions." As Freedman interpreted the scan results, the voters were "mentally battling their attraction to the other side." Not only that, but political affiliations may not be "driven by a deep commitment to issues" but instead relate more to long-term allegiances, similar to loyalty to a sports team.

Before getting to the neuroscience of the study, it's useful to spend a few moments pondering the underlying politics in order to get a feeling for how in the neurosociety brain science can be coopted in the interest of personal agendas. Joshua Freedman is the brother of Tom Freedman, a strategist in the 1996 Clinton campaign. Along with William Knapp, a consultant for the 1992 and 1996 Clinton campaigns and the 2000 Gore campaign, the Freedmans founded the company (FKF Research) that sponsored the UCLA study.

When questioned about this strange alliance between a neuroscientific institute and politics, Freedman told science journalist

Lisa Phillips that FKF aims to research potential applications for fMRI for political and media consulting. Nothing had been said earlier in his bio for the op-ed piece about Freedman's interest in political and media consulting, nor was he identified as a co-founder of the company that sponsored the research.

On the neuroscience side, the fish doesn't smell any better. At the time of Freedman's op-ed piece the fMRI study had yet to be published. Nor had it been subjected to the usual peer review process whereby experts in the field comment on the scientific validity of a research study. But comments on the study came aplenty after Freedman called attention to it.

Kenneth Heilman, a University of Florida neurologist and international authority on behavioral neurology, called the study "physiological phrenology with a whole bunch of psychobabble." As Heilman observed, selective activation of a brain area is open to many different interpretations. While it's true that the anterior cingulate may be activated during conflict, it also springs into action when people are concentrating and focusing their attention, attempting to rationally review several possible alternatives, or experiencing difficulty in processing information.

In addition, Heilman and others were disturbed by the lack of controls in the UCLA study, which claimed a linkage between the brain responses occurring in only twenty people with those in the brains of the nation's millions of voters. "You need control procedures and if you don't have them all that you end up with is a physiological phrenology," concluded Heilman.

Aside from the possible political motivation of both the research and the subsequent op-ed piece, the implicit assumptions about brain function and political persuasion are deeply disturbing. Certainly more is being promised here than can ever be delivered. But that may not stop some political consultants from looking to brain science to provide formulas for winning campaigns. Here's one form that manipulation might take.

My Brain Made Me Vote for Her

Take out a full-face picture of yourself and stare at it for a moment. Place beside it a picture of a politician (of the same gender, to make this a bit easier). Now imagine that I digitally alter the politician's picture to produce a blend that contains 40 percent of your picture and 60 percent of the politician's features. The resulting morphed picture still looks like the politician but with elements drawn from your picture. What effect do you think this would have on your attitude toward the politician? This experiment has actually been carried out. Most people evaluated the morphed politician more favorably than the unaltered photograph. Somehow they detected that the politician's picture included features from their own face.

"Given these results suggesting the powerful effects of similarity on interpersonal attraction, it is not implausible to suggest that image manipulation may become a popular (albeit arguably immoral) strategy for campaign strategists," suggest the authors of the study. They predict an increased use of morphed faces in campaign advertising. Although this may initially sound rather far out, the technology for morphing faces already exists thanks to software packages that can automatically detect and manipulate selected facial features.

Thus a picture of you taken from a driver's license could be subtly morphed with a candidate's campaign photo to produce a composite that would appeal to you as a result of your unconscious identification with the morphed picture. Still unresolved is the question of how you would respond if 30 percent or 20 percent or perhaps only 10 percent of your features appear in the morphed picture. It's conceivable that only a small percentage of your features need be incorporated for you to respond favorably. If the percentage is small enough—5 percent or less—it would be unlikely that anyone, including you, would consciously notice the difference.

If you doubt the feasibility of using morphed photos, consider the fact that photographic databases are developing in tandem with nationwide demographic databases tailored to geographical districts and zip codes. For example, catalogs and marketing announcements are mailed out according to zip codes, which in most cases correspond to targets such as the recipient's personal income and social status. It would not be all that complicated to send you a morphed picture of a political candidate incorporating elements from your own face drawn from, say, your driver's license photo. Would your viewing of that morphed picture be sufficient to swing your vote toward that candidate? At the moment nobody can answer that question because nothing like this has ever been tried. But the authors of the study issue this caution: "In elections where voters possess little if any substantive information, they might have no choice but to rely on visual clues. And for voters who have attributes in common with the candidate, facial similarity is an important clue." They further warn, "Software that automatically morphs a candidate's image based on stock images that are archived with a voter's address is a possibility that could arrive in months rather than years."

If it turns out to be impractical to morph candidates' pictures in order to make them look similar to individual voters, then simply making a subtle change in a candidate's facial expression may turn out to be an equally powerful incentive to prefer that candidate. In a study testing this, Alexander Todorov asked some college undergraduates to look for only one second at pictures of the faces of past candidates for the United States Senate and House of Representatives. Students who knew how the election had turned out were eliminated from the study. The remaining volunteers were then asked to choose which candidate appeared more competent.

Todorov selected competence because of the emphasis voters place on that trait when deciding whom to vote for. The results of the study confirmed this: Despite the short amount of time during which the students viewed the pictures, they selected the winning candidate with an accuracy of about 70 percent. The experiment was then repeated, and this time the participants were asked to make judgments about intelligence, leadership, honesty, trustworthiness, charisma, and likability. None of these qualities led to the successful selection of those politicians who had won their elections. Only judgments about competence predicted the outcomes of the elections.

Todorov's findings suggest that inferences about other people based on facial appearance can override slower, more "rational" appraisals. "Person impressions can be formed 'on-line' in the very first encounter with the person and can have subtle and often subjectively unrecognized effects on subsequent deliberate judgments," writes Todorov.

Of course, there are limits to the possible political contributions of social neuroscience within the neurosociety. You won't be able to distinguish liberals from conservatives by studying their fMRIs. That determination still needs to be made on the basis of voting records and responses to certain hot-button issues. And while it's true that an fMRI may pick up some indications of conflict in a person's political opinions, as in the UCLA study, why should that be surprising? When it comes to politically charged issues, few of us—whatever our politics—are completely satisfied with the status quo. So why should anyone be surprised if fMRI scans detect activity patterns reflecting our dissatisfaction and ambivalence?

But more is at stake here than simply an overconfidence on the part of some researchers that people's politics can be read by brain scans, or that a person might be enticed to vote for a candidate with perceived similar facial features or a face that suggests competence. Rather, the danger involves the misuse of legitimate social neuroscience by entrepreneurial and self-promoting interests, as with the UCLA study. Think of that study the next time you read or hear about someone claiming to have discovered a "brain center" responsible for a specific trait or behavior. Such a claim is pure hooey. While it's true that a lot of correlations can be established between brain and behavior, many of them described in this book, the brain is too highly interconnected for any simplistic one-to-one correlation between a single brain area and our complex and nuanced human faculties.

The more sophisticated the function, the less likely it can be pinned down to a specific localization. Since the brain is organized in circuits rather than centers, you can give up the idea that an fMRI or some future technology is going to help you decide whether you're "really" a Republican or a Democrat or which of several candidates you should vote for. It just isn't going to happen.

Afterword

Future Directions

While the human brain hasn't changed in size during the last 200,000 years, recent genetic research suggests that other features of the brain are still evolving. In late 2005 researchers discovered variants in two genes that control brain development, which raises the intriguing possibility that these new variants may endow their carriers with cognitive advantages such as higher intelligence or quicker thinking. Although additional observation and research will be required to prove that, the discovery of those two genes suggests that we should stop thinking of the brain as finalized.

Further, we now know that culture exerts a more powerful effect than strictly biological factors in shaping our brains. Take, for example, the dramatic increase in the average IQ, which has climbed 24 points since 1918. Is this increase due to better education? Healthier diet? The greater amounts of information that our brains must now process compared to only a few years ago?

Whichever explanation you favor (I favor the information link), the boost in IQ didn't come about simply by natural-selection-driven changes in the brain: The time frame is simply too short for such a degree of brain modification to be wrought by biological influences alone.

During the remainder of the twenty-first century we can expect technology to continue to exert a profound influence on our brain's development. According to Ray Kurzweil, author of *The Singularity Is Near: When Humans Transcend Biology*, "We won't experience 100 years of technological advance in the 21st century; we will witness in the order of 20,000 years of progress when measured by the rate of progress in 2000, or about 1000 times that in the 20th century."

Moreover, the most powerful effect on our brain will involve information technology, which is already altering our attitudes, our ways of relating to each other, the speed and cadence of our speech, and even our attitudes toward work and leisure.

Coupled with this insight that culture, especially information technology, is now the main driver of evolutionary development comes a second insight: The human brain, the organ of the mind, cannot be understood by focusing on single individuals. Since we are social creatures, a need to belong is as basic to our survival as our need for food and oxygen. Indeed, attachment and nurturing associations with other people play important roles in our lives from the moment of birth.

As a result of our dependence on other people, evolution has genetically programmed into us an exquisite sensitivity to each other's social signals. That's why exclusion, ridicule, separation, divorce, and bereavement hurt so much and exert such negative effects on us. Conversely, attraction, altruism, identification, communication, compassion, and cooperation exert positive effects on our brain by making us feel that we belong.

Neuroethics: Is This Life Worth Living?

Increasingly during the twenty-first century we will be encouraged to think of ourselves in terms and concepts drawn from neuroscience. For example, take the much-publicized conflict in the spring of 2005 about whether Terri Schiavo should have her feeding tube reinserted.

During the several weeks over which this drama played out, neurologists and ethicists flitted from one talk show to another, explaining with sometimes impassioned rhetoric why a person in a persistent vegetative state could not survive. At some point they inevitably segued into a discussion of whether Terri Schiavo was "still a person" and whether her life was worth living. To bolster their point, they displayed MRIs of Schiavo's brain, which contained huge, gaping holes resulting from extensive loss of brain substance.

After viewing the images on television, some people began to change their minds about her chances for recovery. Still later, when electroencephalograms and other tests also pointed to a persistent vegetative state, the only remaining people insisting on the possibility of Schiavo's survival were those driven by ideology or politics. Throughout all of this, technological tests measuring brain function shaped public opinion and largely determined the woman's fate. Personhood, it seems, demands a certain level of neurological activity, which Terri Schiavo lacked.

Incidentally, I'm in complete agreement that Terri Schiavo was in a terminal state. In this sensitive and heart-wrenching situation, I think the correct decision was made: that Schiavo had no chance for recovery. I'm making a different point: Brain scans are being increasingly used to determine whether life is "meaningful." The question has now switched from "Is the patient dead or alive?" to "Is this life worth living?" This latter question isn't really something that can be decided purely on the basis of brain scans.

Confident (I would argue that they are overconfident) assertions are being made about the value of people's lives based on brain scans. These claims ignore or underplay many of the uncertainties about these technologies. "These images are quite seductive," says Marcus Raichle, coauthor of *Images of Mind* and a pioneer in the use of brain imaging. "It's intuitively easy to relate to a picture, and that's both good and bad."

Pictures speak louder than words, as we've heard all of our lives. But for every picture there must be an interpretation—and there's the rub. I'm thinking now of a CAT scan we used several years ago in the filming of a segment for the PBS television production of *The Brain*. It showed such extensive cortical atrophy that the brain consisted of nothing more than a thin ribbon of tissue—what looked like a vast empty chasm. We invited the viewers to speculate about the mental state of the person with this profoundly abnormal CAT scan.

Few people guessed that the subject of that CAT scan was a perfectly normal man who had finished high school, married, and worked as a bank teller. Why did the CAT scan in this instance prove so useless as a predictor of mental functioning? The key concept—one that couldn't be conveyed in the picture—was that the CAT scan had been made when this man was an infant and that somehow, seemingly miraculously, his brain had compensated for the extensive loss of substance. The image, in short, had to be put in context: The immature brain possesses recuperative powers far exceeding anything that occurs in adults.

Context is always important when drawing conclusions from brain images. An fMRI doesn't provide actual pictures of what's going on in the brain. Rather, the images result from extensive mathematical processing of raw data. What's more, that processing is carried out in ways chosen by the researchers based on their interests and preconceptions. In addition, brain images taken for research projects are generally composites drawn from many people, not from a single individual. And even on those occasions when the images are drawn from one person, they don't correspond to brain activity at a specific moment but, instead, represent data gathered during many repetitions of the same test. Further, what qualifies as an "abnormality" in a given scan depends on databases and statistical thresholds (depicted by the color changes on the scan). If the databases or thresholds are altered, what initially appeared as an overly active area may fade into the general background of the image.

Context is also important whenever we attempt to link human traits and performances with specific locations within the brain. "People yearn to find 'the' neural correlate of an emotion or even a creative thought. That hope seems about as silly as trying to find the key to the greatness of a novel by closely examining the typographical symbols that compose it," as cognitive science researcher Douglas Hofstadter phrased the dilemma in an interview for a symposium on creativity sponsored by *New Scientist*.

Therefore we should be cautious about defining ourselves strictly in terms of the brain. If we fail to do this, we face a series of decisions that brain science cannot resolve. For instance, what attitude should we take toward an advanced Alzheimer's patient who lacks awareness of people, places, and events? Should we consider his life as no longer valuable? What attitude should we take toward the patient in a coma who isn't likely to recover but for whom a slim possibility for recovery still exists? Brain science can provide some help in deciding these questions, but it shouldn't be the final arbiter.

While we can learn a lot about ourselves by learning more about the brain—indeed, that has been the message of my previous seventeen books—there are limits when it comes to translating mind matters into brain matters. Undoubtedly, brain science has much to contribute to our lives. But it cannot be allowed to define us.

Make Me Smarter

As we learn more about the brain, the emphasis will shift from seeking treatments for brain diseases toward developing chemical and other techniques to enhance mental performance in the normal brain.

For instance, while everybody would like to be smarter, drugs aimed at bumping our IQ up a few points have only recently moved from fantasy to reality. At the forefront are memory-enhancing drugs, presently one of the most important targets of the pharmaceutical companies. (I detail some of this research on drug development in memory in my earlier book *The New Brain*.) But rather than discussing the research that has been carried out since then, let's concentrate on the core issue: What brain function would you most like to improve?

Focus, concentration, memory, and mental endurance would lead my personal list. Why these four? Because most of our intellectual achievements involve one or more of these processes. Focus and concentration help us to remain centered when engaged in mental challenges. Memory helps us to maintain the requisite mental linkages between earlier knowledge and current learning. Mental endurance is perhaps a less familiar concept best illustrated by an analogy. Just as our physical performance declines whenever our bodies become fatigued, we can also become mentally fatigued and give in to distraction—we mentally "veg out" and redirect our mental energies to less demanding tasks.

Drugs capable of strengthening the four mental attributes have been available for years. Amphetamine and other psycho-stimulants such as Ritalin can enhance focus, concentration, memory, and mental endurance. Indeed, such psychostimulants have long been a frequent refuge for unprepared students who prior to exams are forced to learn large amounts of class material in a short period of time. But doctors today rarely prescribe psychostimulants as cognitive enhancers. For one thing, the drugs are tightly controlled by the FDA and can only be prescribed for specific conditions such as attention deficit disorder. Even more important, psychostimulants can produce a lot of undesirable side effects, including the need to take increasing amounts of the drug in order to get the same effect each time.

Despite the potentially life-threatening risks associated with psychostimulants, the number of students using the drugs (often obtained on the street) has reached an all-time high. Prescriptions for the drugs for adults age twenty to thirty tripled from 2000 to 2004. As a measure of less legitimate channels, 14 percent of students at a midwestern liberal arts college admitted to buying or borrowing prescription stimulants from each other; an additional 44 percent said they knew someone who obtained the drugs through illegal means. Another informal survey taken at an Ivy League college places the number of users at closer to 60 percent.

The use of psychostimulants is likely to become even more common thanks to the recent availability of a new generation of alertness-enhancing drugs that are largely free of side effects and can improve performance in those four key areas, especially mental endurance. Although officially approved only for narcolepsy (a rare disorder marked by episodes of irresistible drowsiness and sometimes falling asleep under inappropriate circumstances) and shift-work-induced drowsiness, Provigil (modafinil) is now routinely used by businesspeople, entertainers, military personnel, and others to remain fully awake for long periods, sometimes even days at a time. Thanks to Provigil and similar drugs currently under development, normal fatigue cycles will no longer determine an individual's capacity for sustained effort. Which leads to a pivotal question: Should these drugs be controlled along lines similar to those currently applied to steroids among athletes?

Answering yes to that question is intuitively appealing. Why should one student achieve superior scores on a standardized test compared to his equally knowledgeable peers simply because prior to the test he swallowed pills that enhanced his focus, concentration, memory, and mental endurance?

Others, taking a longer view, might argue that scores on a test aren't predictive of future achievement: genius is 1 percent inspiration and 99 percent perspiration, as Thomas Edison put it. Even if the medicated student scores higher on that standardized test and thus is judged "smarter" than other equally qualified students, "smart is only part of the story," asserts neuroscientist Michael Gazzaniga. "Smart describes how well one processes information and figures out tasks. Once something has been figured out, much work must be applied to the solution, and the smartest people in the world rarely say that the work applied to solutions is easy."

Personally, I have to admit that I take the shorter view. In the interest of maintaining a level playing field, I think medications should be tightly controlled in such competitive situations as standardized nationwide exams. Monitoring of drug use might even require some form of drug testing prior to nationwide competitive exams. (Tests of saliva would likely replace urine and blood tests, which are impractical in an examination situation.) Students are already required to show proof of identity prior to the exams, so why not also require proof that their performance on the exam isn't going to be influenced by cognition-enhancing drugs? (I'm exempting from the drug testing students who are prescribed cognitive enhancers for conditions such as severe ADHD.)

But the use of "smart drugs" and "designer drugs" won't be limited to students taking competitive exams. Nor will the use of drugs remain restricted to raising one's IQ, combating normal fa-

tigue cycles, or other personality enhancements. Mood modification will be an equally important goal.

Thanks to ongoing worldwide research on the brain, we can
expect performance-enhancing neurotechnology to influence
some of our most basic human emotions and experiences such as
grief, fatigue, lack of focus, mood variations, forgetfulness, distraction, and other "unproductive" mental states. For example, consider the millions of Americans currently taking tranquilizers and
antidepressants. While the large majority of them are suffering
from anxiety and depression, others are taking the drugs in order
to bring about lifestyle and personality changes. People who would
traditionally be described as shy are now treated with drugs aimed
at "curing" the questionable and recently defined "disorder" labeled as social phobia. Undoubtedly, extreme examples of this
condition are clearly pathological (such as an unwillingness to
leave one's apartment in order to avoid contact with other people), but what about a person who functions well enough in social
situations but, given his choice, prefers the company of familiar
friends? Is he just shy? Or is he a social phobic?

A similar misdiagnosis of a normal personality variant occurs
in many people currently labeled as dysthymic. Think of someone
you know with a wry and dour personality who's given to pessimistic expressions about just about everything. Even though
someone like that isn't always a joy to be around, most such people aren't really clinically depressed, deny that they're depressed
when asked, and function passably well at home and at work.
Nonetheless, they are encouraged by means of TV advertisements
to think of themselves as depressed and in need of antidepressant drugs.

Powerful drugs are already available that are capable of helping
people to work longer and smarter. How should an employee
respond if his boss encourages him to take a drug that might

improve his job performance but may later turn out to be bad for his health? Even if only a small percentage of workers decide to take mental-performance-enhancing drugs, the ultimate effect on company morale could be devastating. Would you take a medicine that would enable you to work longer hours without loss of efficiency? If the answer is no, would you feel differently if you knew several of your associates were more than willing to take the drug in order to snag your job?

Certainly, it will be difficult to compete with a coworker taking a fatigue-eliminating drug such as Provigil, another drug capable of enhancing memory and concentration, and a third drug that stabilizes mood and combats the depression that results from an all-work-and-no-play lifestyle. Given the fierce competition existing in our society, employees will likely feel subtle and not-so-subtle pressures to use psychoactive drugs to enhance their performance—to become, in essence, "office-obsessed" employees.

To bioethicist Thomas H. Murray, a consultant to the Olympic committee on the ethical implications of drug enhancement in sports, there is no simple answer to the use of cognitive enhancers in schools or the workplace. "Cognitive enhancers could increase everyone's cognitive abilities. What would be so bad about that? Is taking cognitive enhancers just another practice of good educational or mental health, like providing a positive learning environment and practicing good study habits? Is popping a pill a 'quick fix' instead of working toward improving cognitive ability through reading?"

Further complicating the picture are new forms of neuro-social engineering aimed at altering perfectly normal human emotions, such as grief, for which our society has limited tolerance. Has a loved one just died? Attend the funeral, of course, and perhaps take a day off from work, but then return to the office. And when you return, don't mope or demonstrate any other morbid responses that may affect your own productivity or the productivity of those around you. Indeed, why not streamline the whole process by taking a tranquilizer prior to the funeral and an antidepressant during the weeks following? If you can't sleep, there are sleeping pills to be taken before bed, followed by alertness-enhancing drugs in the morning to bring you right up to speed. From the point of view of both the employee and the employer, performance-enhancing drugs represent such chemical panaceas: The employee works through his or her grief reaction with little variation in productivity.

In the twenty-first-century workplace it is only too easy to forget that chemically modifying our thoughts, mood, and behavior by means of mind-altering drugs may not necessarily be in our best long-term interest.

The Brain of an Extrovert

Over the next two decades we can also expect social neuroscience to shed additional light on how our perceptions and processing of social signals influence our behavior. Indeed, brain-based insights into everyday social interactions will become commonplace.

For example, personality traits such as extroversion can already be correlated with distinct brain responses. When an extrovert encounters a positive word such as *lucky*, his brain's response is clearly different than the response to a negative term such as *cancer*, according to neuroscientist B. W. Haas of the State University of New York at Stony Brook. While processing the positive word, the extrovert's brain shows increased activation in the anterior cingulate. This correlation is sufficiently robust that a trained observer such as Haas can estimate a person's tendency toward extroversion simply by observing via fMRI his or her brain's anterior cingulate response to positive and negative words. It's as if the extrovert's brain is drawn to positive words and lingers on them a few milliseconds longer.

Behaviorally, a similar process takes place as well: Extroverts tend to dwell on the positive and de-emphasize the negative—one of the reasons we enjoy spending time with an extrovert. As a practical application, such a test when administered as part of a pre-employment assessment could provide a good measure of a person's tendency toward extroversion—a highly desirable trait in someone who is being hired for sales or other jobs that involve a lot of social interaction and team effort.

A specific brain response also occurs when socially adept people reason about social situations, such as estimating whether another person can be trusted. Individuals with high social intelligence quickly arrive at such decisions, while the less socially adept take longer. Although the same brain area (the prefrontal cortex) activates in both instances, the brain of the person with higher social intelligence shows less activity in that area than his counterpart with lower social intelligence.

"This suggests that people with higher social–emotional intelligence may have an easier time reasoning about social situations," according to Deidre Kolarick Reis, who headed the Yale University team that carried out the fMRI study. "Individuals with higher social intelligence may be more efficient, requiring less 'brainpower' to solve social reasoning problems."

In other words, the more adept you are at reading social clues, the less effort your prefrontal cortex has to expend. In practical terms, an employer could get some measure of a job applicant's social skills by measuring the person's prefrontal activation patterns while she answered questions about how she would manage interpersonally challenging social situations. The availability of such technology creates a dilemma similar to the cognitive and mood enhancers: The employee who refuses the tests will be at a distinct disadvantage compared to the employee who is willing to be tested.

One career application of the new neurotechnology already in place involves the selection of appropriate personnel for Special Forces (SF), an elite military group trained to remain emotionally resilient under conditions that would induce severe psychological distress in most people. In order to perform his duties, an SF soldier must be able to keep his emotions under control—especially feelings of anger, horror, and fear. Military commanders have long wished for an objective way to predict a soldier's emotional resiliency prior to assigning him to the SF. Unfortunately, there hasn't been any way of knowing for certain a particular SF soldier's responses until he's actually in the field. Now prediction of emotional resiliency is possible, according to fMRI research carried out at the National Institute of Mental Health (NIMH).

Special Forces soldiers looked at a series of faces shown to them while lying in an fMRI scanner. The ostensible task was for them to identify the gender of the people in the pictures. In fact, the test involved measuring the soldiers' responses to those faces in the series that depicted anger or fear.

In comparison to untrained volunteers, the SF soldiers showed increased activation in two areas (the anterior cingulate and the inferior frontal cortex) known to be associated with emotional regulation. At the same time the SF soldiers also showed heightened responses within the amygdala, the center most closely associated with fear and general emotional arousal. Thus the SF soldiers process fear just like the rest of us but differ from the untrained volunteers in their ability to retain firm control over that fear thanks to enhanced activity in the brain areas that regulate fear. "Individuals who are resilient to trauma may possess a unique psychobiology that either enables them to modulate fear or to reappraise their traumatic experiences in a positive way," according to Marina Nakic, who coordinated the study.

Although the SF study is a small one (only thirteen soldiers were tested), consider the implications: The optimal candidates for SF and other high-risk assignments might now be identified at an early stage in the selection process. Thus the subjective factor in selecting the best men for inclusion in the SF and other emotionally demanding work may soon be greatly reduced. While that is a positive application of neuroscience, I'm less enthusiastic about "brain tests" aimed at screening people for less sensitive and emotionally demanding occupations.

Within the neurosociety, social neuroscience will also change many of our ideas about maladaptive behaviors such as addiction. Until recently addictions were thought to be restricted to chemicals: alcohol, crack cocaine, nicotine, and other legal and illegal substances. A unitary theory of addictions evolved over the past decade based on the brain's pleasure circuits and the neurotransmitter dopamine. Recently, a similar dopamine-mediated stimulation of the pleasure circuits has been discovered to underlie behaviors such as computer gaming and gambling.

Sabine Grusser-Sinopoli, a researcher at Charite-Universite Medicine in Berlin, has found that people who feel compelled to spend hours engaged in computer games experience both arousal (leading to excessive computer gaming) and craving (leading to engaging in computer gaming to the neglect of other aspects of their lives). Both arousal and craving can be measured physiologically and in excessive computer gamers "are similar to what we found in drug users," Grusser-Sinopoli told me.

A similar addiction pattern exists for excessive gambling. According to the research of Jakob Linnet, of Aarhus University in Denmark, high sensation seekers are more likely to become addicted to excitement-inducing games such as poker or blackjack. A greater release of dopamine is also characteristic of these high sensation seekers. "Pathological gamblers feel a kick or rush from gambling, despite the negative consequences of losing money," says Linnet.

Questions and Challenges

Even though we have learned a lot about the brain mechanisms underlying social interaction, questions still remain.

Can specific mental abilities be genetically selected? Sarah-Jayne Blakemore of the Institute of Cognitive Neuroscience at University College London points to our pet dogs as one example of what can take place when social patterns modify brain structure and resulting behavior. If you own a dog, you know the value of keeping your eyes averted when the dog wants to go for a walk but you don't. One careless look in the direction of the front door or the leash, and your dog starts barking to go out. That happens because the dog follows your gaze and knows exactly what you're looking at. It then uses this information to try to get you to turn off the TV and go for that walk. And while neither wolves nor dogs raised in the wild can use a person's eye gaze to infer intention, puppies monitor human gaze information almost from birth. This suggests that eye-gaze monitoring has been bred into our pet dogs' gene pool from centuries of domestication. This use of eye-gaze information by dogs evolved naturally in response to cultural and social patterns that led to keeping dogs as pets.

A similar calculus applies to us. Social interaction modifies both brain function and, given enough time, genetic selection. Will future modifications continue to mirror our patterns of social interaction? If so, what will our brains be like when, as futurist Ray Kurzweil predicts, "ultimately we will merge with our technology"? "As we get to the 2030s, the non-biological portion of our intelligence such as computers and prosthetic devices will predominate.

By the 2040s it will be billions of times more capable than the biological portion," according to Kurzweil.

Why does the observation of another person's behavior exert such a powerful effect on our own? A similar influence doesn't occur if we watch a robot performing the same movements. As a corollary, why is it that staring intently at someone or hearing someone call our name activates portions of our brain known to be involved in our understanding of others?

Does a common mechanism underlie our ability to attribute mental states to ourselves as well as other people? If so, we may be able to use this knowledge to enhance self-understanding as well as facilitate communication between individuals and between nations. Putting oneself in another person's shoes may soon become a brain-based metaphor.

What should be our responses as brain imaging and other technologies reveal aspects of ourselves that we may not wish to reveal or in some cases may not even be aware of ourselves? For example, recent research shows that in obese women the striatum, a part of the brain's reward system, is selectively activated when the women look at highly caloric food, or even just a picture of it. Since this response doesn't occur in women of normal weight, some neuroscientists are suggesting that obese women can be compared to drug addicts, who display similar striatal activation when looking at pictures of drug paraphernalia.

"Obesity and addiction are special cases of the consequences of ingestive behavior gone awry," says Nora Volkow, director of the National Institute on Drug Abuse. Although additional research on the linkage between drug addiction and obesity may provide a novel and potentially useful approach to obesity, such a treatment path will at least initially run up against a conceptual hurdle: Most obese people don't think of themselves as addicts and, understandably, don't want other people thinking of them that way either.

What limits can and should be placed on marketing and entertainment industry efforts to use insights about the brain in order to influence purchasing? Thanks to brain imaging, it's now possible to design marketing strategies based on measurements of activity taking place within the brain of the consumer. Although I believe there are limits on how much can be accomplished by these efforts, I also believe that neuromarketing, as a manipulative science, will continue to use emerging new knowledge about social neuroscience to help devise strategies aimed at prompting people to act in ways that further the interests of the marketers. And since we have no choice about being exposed to the new neuromarketing techniques (unless we're living in a cave), our best chance for resisting manipulation is to learn as much as possible about the emerging social applications of brain science.

Here's our challenge: We can employ this emerging new knowledge about social neuroscience to advance human freedom within the neurosociety, or we can allow irresponsible people to use this knowledge in ways that are not always to our advantage. By learning as much as we can about the social applications of emerging knowledge about the brain, we will be in a position to resist manipulation by ads, pop culture, political spin, movies, and television. To this extent, social neuroscience can provide us a path toward the achievement of both personal and collective liberation.

Notes

Introduction: Welcome to the Neurosociety

Kevles's comment are from *Science*, June 4, 2004, and an e-mail to the author, June 30, 2005.

The historical material is from John T. Cacioppo, ed., *Foundations in Social Neuroscience* (Cambridge, Mass.: MIT Press, 2002), especially John T. Cacioppo and Gary G. Bernston, "Social Neuroscience."

Chapter 1: The Emergence of the Neurosociety

The Angelo Mosso story is from Michael I. Posner and Marcus E. Raichle, *Images of Mind* (New York: Scientific American Library, 1994), 54–55.

The definition of cognition is taken from Dale Purves et al., eds., *Neuroscience*, 1st ed. (Sunderland, Mass.: Sinauer Associates, 1997), G-3.

The frontal lobe description is from various sources, including Ian Glynn, *An Anatomy of Thought* (Oxford, UK: Oxford University Press, 1999), Chapter 21, "Planning and Attention," and Harvey S. Levin, Howard M. Eisenberg, and Arthur L. Benton, eds., *Frontal Lobe Function and Dysfunction* (New York: Oxford University Press, 1991).

The discussion of theory of mind is from Donald T. Stuss and Robert Knight, eds., *Principles of Frontal Lobe Function* (Oxford, UK: Oxford University Press, 2002).

Chapter 2: How the Brain Processes Information

The Michael Faraday story is from Daniel Wegner, "The Mind's Best Trick: How We Experience Conscious Will," *Trends in Cognitive Sciences* 7, 2 (2003): 65–69.

The Claparède anecdote is from É. Claparède, "Recognition and 'Me-Ness,' " in David Rapaport, ed. and trans., *Organization and Pathology of Thought* (New York: Columbia University Press, 1951).

The card, the Chinese ideograph, and other experiments on unconscious perception are from Philip M. Merikle, "Psychological Investigations of Unconscious Perception," *Journal of Consciousness Studies* 5, 1 (1998), and Philip M. Merikle and Meredyth Daneman, "Conscious vs. Unconscious Perception," in Michael Gazzaniga, ed., *The New Cognitive Neuroscience* (Cambridge, Mass.: MIT Press, 2000).

John Bargh's comments are from his "Bypassing the Will: Toward Demystifying the Nonconscious Control of Social Behavior," in Ran R. Hassin, James S. Uleman, and John A. Bargh, eds., *The New Unconscious* (Oxford, UK: Oxford University Press, 2005).

Ap Dijksterhuis's comments are from Ap Dijksterhuis, Henk Aarts, and Pamela K. Smith, "The Power of the Subliminal: On Subliminal Persuasion and Other Potential Applications," in Hassin, Uleman, and Bargh, eds., *The New Unconscious*, 86.

The Wilder Penfield quotes are from his book *The Mystery of the Mind* (Princeton, N.J.: Princeton University Press, 1975), 21, 77.

The Roger Sperry quote is from Kenneth M. Heilman and Edward Valenstein, *Clinical Neuropsychology*, 1st ed. (New York: Oxford University Press, 1979), 323.

"A million connecting fibers" is from Mark R. Rosenzweig, S. Marc Breedlove, and Arnold L. Leiman, *Biological Psychology: An Introduction to Behavioral, Cognitive, and Clinical Neuroscience*, 3rd ed. (Sunderland, Mass.: Sinauer Associates, 2002), 624.

The Gazzaniga experiments are from John T. Cacioppo, ed., *Foundations in Social Neuroscience* (Cambridge, Mass.: MIT Press, 2002), 55.

The Sorenson quote is from his book *A Brief History of the Paradox: Philosophy and the Labyrinths of the Mind* (New York: Oxford University Press, 2003), 186.

The Libet experiment is described in Benjamin Libet, "Conscious vs. Neural Time," *Nature* 352, 6630 (1991): 27–28.

Rocky is described in Heilman and Valenstein, *Clinical Neuropsychology*, 334.

The François Lhermitte observations are from Richard M. Restak, *The Mind* (New York: Bantam, 1988), 268–71.

Chapter 3: The Emotional Brain

An excellent description of the "high road" and the "low road" of the fear pathway appears in Joseph LeDoux, *The Emotional Brain: The Mysterious Underpinnings of Emotional Life* (New York, Simon & Schuster, 1996).

The material about wide-eyed expressions is taken from Paul Whalen et al., "Human Amygdala Responsivity to Masked Fearful Eye Whites," *Science*, December 17, 2004, 2061.

Nalini Ambady and gaze direction are from Reginald B. Adams Jr. et al., "Effects of Gaze on Amygdala Sensitivity to Anger and Fear Faces," *Science*, June 6, 2003, 1536.

Attentional blink is from Elizabeth A. Phelps, "The Interaction of Emotion and Cognition: The Relation Between the Human Amygdala and Cognitive Awareness," in Ran R. Hassin, James S. Uleman, and John A. Bargh, eds., *The New Unconscious* (Oxford, UK: Oxford University Press, 2005), 71–72.

Chapter 4: How Our Brain Constructs Our Mental World

The Helmholtz story is from George R. Mangun, "Neuronal Mechanisms of Attention," in Alberto Zani and Alice Mado Proverbio, eds., *The Cognitive Electrophysiology of Mind and Brain* (Boston: Academic Press, 2003).

The experiments on arm positions and preferences is taken from John A. Bargh and Tanya L. Chartrand, "The Unbearable Automaticity of Being," *American Psychologist*, July 1999, 475.

The Bargh research is described in ibid.

The William James quotes are from his *Principles of Psychology* (New York: Holt, 1890).

The monkeys taught to rip paper are from Sarah-Jayne Blakemore, Joel Winston, and Uta Frith, "Social Cognitive Neuroscience: Where Are We Heading?" *Trends in Cognitive Sciences* 8, 5 (2004): 216–22.

Iacoboni's research is described in Greg Miller, "Reflecting on Another's Mind," *Science*, May 13, 2005, 945–47.

Gazzola's research is described in "Hearing What You Are Doing—an fMRI Study of Auditory Empathy," poster B299 presented at the 35th annual meeting of the Society for Neuroscience, Washington, D.C., November 12–16, 2005.

The verb-behavior link is from John A. Bargh, "Bypassing the Will: Toward Demystifying the Nonconscious Control of Social Behavior," in Ran R. Hassin, James S. Uleman, and John A. Bargh, eds., *The New Unconscious* (Oxford, UK: Oxford University Press, 2005).

The mental image research is from Vittorio Gallese, "The Manifold Nature of Interpersonal Relations: The Quest for a Common Mechanism," in Christopher D. Frith and Daniel M. Wolpert, eds., *The Neuroscience of Social Interaction: Decoding, Imitating, and Influencing the Actions of Others* (Oxford, UK: Oxford University Press, 2004).

The Calvo-Merino material is from B. Calvo-Merino et al., "Actions Observation and Acquired Motor Skills: An fMRI Study with Expert Dancers," *Cerebral Cortex* 15 (2005): 1243–49.

The Johansson experiment is described in G. Johansson, "Visual Perception of Biological Motion and a Model of Its Analysis," *Perception and Psychophysics* 14 (1973): 201–211.

A discussion of the importance of the superior temporal sulcus (STS) and the two visual streams can be found in Emily D. Grossman and Randolph Blake, "Brain Areas Active During Visual Perception of Biological Motion," in John T. Cacioppo and Gary G. Berntson, eds., *Social Neuroscience* (New York: Psychology Press, 2004).

A fuller description and discussion of the five diagrams can be found in Fulvio Castelli, Francesca Happé, Chris Frith, and Uta Frith, "Movement and Mind: A Functional Imaging Study of Perception and Interpretation of Complex Intentional Movement Patterns," in Cacioppo and Berntson, eds., *Social Neuroscience*.

The Frith summary is from Castelli, Happé, Frith, and Frith, "Movement and Mind," in Cacioppo and Berntson, eds., *Social Neuroscience*, 166.

The Sally-Ann test is discussed in more detail in Ralph Adolphs, "Cognitive Neuroscience of Human Social Behavior," *Nature Reviews/ Neuroscience* 4 (2003): 165–78.

The Kevin Ochsner comments are from Kevin Ochsner et al., "Reflecting upon Feelings: An fMRI Study of Neural Systems Supporting the Attribution of Emotion to Self and Other," *Journal of Cognitive Neuroscience* 16, 10 (2004): 1746–72. The bank robber and burglar quotes are from Uta Frith and Christopher D. Frith, "Development and Neurophysiology of Mentalizing," in Frith and Wolpert, eds., *The Neuroscience of Social Interaction*, 54.

For additional information on Elizabeth Meins's research, contact Meins at elizabeth.meins@durham.ac.ul or see "Mind-Reading Mums Boost Babies," *New Scientist*, June 4, 2005.

Christopher Frith's comments are from Frith and Frith, "Development and Neurophysiology of Mentalizing," in Frith and Wolpert, eds., *The Neuroscience of Social Interaction*.

An excellent discussion of brain imaging and social neuroscience can be found in Marcus E. Raichle, "Social Neuroscience: A Role for Brain Imaging," *Political Psychology* 24, 4 (2003): 759–64. Comments about the "default mode" can be found in Debra A. Gusnard et al., "Medial Prefrontal Cortex and Self-Referential Mental Activity: Relation to a Default Mode of Brain Function," *Proceedings of the National Academy of Sciences* 98, 7 (2001): 4259–64, available at www.pnas.org/cgi/doi/10.1073/pnas.071043098.

The Japanese dinner example is adapted from Frith and Frith, "Development and Neurophysiology of Mentalizing," in Frith and Wolpert, eds., *The Neuroscience of Social Interaction*.

The F. T. Bacon research is from his "Credibility of Repeated Statements: Memory for Trivia," *Journal of Experimental Psychology: Human Learning and Memory* 5, 3 (1979): 241–52.

The Ian Skurnik material can be found in Ian Skurnik et al., "How Warnings About False Claims Become Recommendations," *Journal of Consumer Research* 31, 4 (2005): 713–24.

The Susan Andersen quote is from Susan M. Andersen, Inga Reznik, and Noah S. Glassman, "The Unconscious Relational Self," in Hassin, Uleman, and Bargh, eds., *The New Unconscious*, 439.

For additional material on the negativity bias, see Tiffany A. Ito et al., "Negative Information Weighs More Heavily on the Brain: The Negativity Bias in Evaluative Categorizations," in John T. Cacioppo, ed., *Foundations in Social Neuroscience* (Cambridge, Mass.: MIT Press, 2002).

Regarding extroverted personalities, see T. Yarkoni and D. Barch, "Individual Differences in Neural Reactivity to Emotional Stimuli," poster 935.12 presented at the 35th annual meeting of the Society for Neuroscience, Washington, D.C., November 12–16, 2005, or contact the first author at tyarkoni@wustl.edu.

The Susan Fiske material is from a personal communication and her talk, "Perils of Prejudice: Emotional Biases in Brain, Mind, and Society," at the University of Washington, Seattle, Wash., November 3, 2004.

The Sammy Davis Jr. quote is from Michael Anderson's review of Wil Haygood's *In Black and White*, in the *Times Literary Supplement*, November 19, 2004.

For an interesting review of the simulated shooting situations, see J. Correll et al., "The Police Officer's Dilemma: Using Ethnicity to Disambiguate Potentially Threatening Individuals," *Journal of Personality and Social Psychology* 83, 6 (2002): 1314–29.

The Elizabeth Phelps quotes and material are from Elizabeth A. Phelps et al., "Performance on Indirect Measures of Race Evaluation Predicts Amygdala Activation," *Foundations in Social Neuroscience* 12, 5 (2000): 729–38.

Chapter 5: The Empathic Brain: Blurring the Boundaries Between Self and Others

The rubber hand illusion was first described in H. Henrik Ehrsson, Charles Spence, and Richard E. Passingham, "That's My Hand! Activity in Premotor Cortex Reflects Feeling of Ownership of a Limb," *Science Express*, July 1, 2004.

The hand-strapped-to-a-machine experiment is from Jean Decety and Philip L. Jackson, "The Functional Architecture of Human Empathy," *Behavioral and Cognitive Neuroscience Reviews* 3, 2 (2004): 71–100.

The woman and boyfriend experiment was first reported in Tania Singer et al., "Empathy for Pain Involves the Affective but Not Sensory Components of Pain," *Science*, February 20, 2004.

The quote from the security service interrogator is from Michael Bond and Michael Koubi, "The Enforcer," *New Scientist*, November 20, 2004, 47.

The Roger Scruton quote is from Roger Scruton, "Flesh from the Butcher," *Times Literary Supplement*, April 15, 2005, 11.

The Rain Carter quote is from *The Sandcastle*, by Iris Murdoch (New York: Penguin, 1957), 45.

Polk's research is from "The Role of Heredity in Cortical Responses to Faces, Places, and Words: An fMRI Study of Identical and Fraternal Twins," poster 819.2 presented at the 34th annual meeting of the Society for Neuroscience, San Diego, Calif., October 23–27, 2004.

Julian Keenan's research is described in Mark R. Rosenzweig, S. Marc Breedlove, and Arnold L. Leiman, *Biological Psychology: An Introduction to Behavioral, Cognitive, and Clinical Neuroscience*, 3rd ed. (Sunderland, Mass.: Sinauer Associates, 2002), 633.

The Marilyn Monroe–Margaret Thatcher experiment is taken from Pia Rotshtein et al., "Morphing Marilyn into Maggie Dissociates Physical and Identity Face Representations in the Brain," *Nature Neuroscience* 8, 1 (2005): 107–13.

The Adolphs quote is from Ralph Adolphs, "Cognitive Neuroscience of Human Social Behavior," *Nature Reviews/Neuroscience* 4 (March 2003): 167.

Emotional contagion is from Decety and Jackson, "The Functional Architecture of Human Empathy."

The mouth movements of mothers and infants research and the quote from Chartrand are from Tanya L. Chartrand, William W. Maddux, and Jessica L. Lakin, "Beyond the Perception-Behavior Link: The Ubiquitous Utility and Motivational Moderators of Nonconscious Mimicry," in Ran R. Hassin, James S. Uleman, and John A. Bargh, eds., *The New Unconscious* (Oxford, UK: Oxford University Press, 2005).

The Vittorio Gallese quote is from Vittorio Gallese, "The Manifold Nature of Interpersonal Relations: The Quest for a Common Mechanism," in Christopher D. Frith and Daniel M. Wolpert, eds., *The Neuroscience of Social Interaction: Decoding, Imitating, and Influencing the Actions of Others* (Oxford, UK: Oxford University Press, 2004), 162–63.

The LaFrance research is described in Michael Dorris, "Posture Mirroring and Rapport," in Martha Davis, ed., *Interaction Rhythms: Periodicity in Communicative Behavior* (New York: Human Sciences Press, 1982), 279–98.

The Ellen Burstyn quote is taken from *On the Sea of Memory*, by Jonathan Cott (New York: Random House, 2005) 135–36.

The Marco Iacoboni experiments are described in Anahad O'Connor, "Brain Scans Substantiate Feel-the-Pain Sentiments," *New York Times*, February 24, 2004.

The Laurie Carr research is described in Laurie Carr et al., "Neural Mechanisms of Empathy in Humans: A Relay from Neural Systems for Imitation to Limbic Areas," in John T. Cacioppo and Gary G. Berntson, eds., *Social Neuroscience* (New York: Ohio State University Psychology Press, 2004).

Loving-kindness-compassion meditation research is in A. Lutz, J. Brefczynski-Lewis, and R. Davidson, "Loving-Kindness and Compassion Meditation Results in Unique Patterns of fMRI Activation and Enhances the Reactivity of the Insula/Cingulate Neural Circuitry to Negative Stimuli in Meditators," poster 715.9 presented at the 34th annual meeting of the Society for Neuroscience, San Diego, Calif., October 23–27, 2004.

The Dalai Lama quote is from his book *Transforming the Mind: Teachings on Generating Compassion* (London: Thorsons, 2000).

The Shamata meditation material is from J. Brefczynski-Lewis, A. Lutz, and R. Davidson, "A Neural Correlate of Attentional Expertise in Long-Time Buddhist Practitioners," poster 715.8 presented at the 34th annual meeting of the Society for Neuroscience, San Diego, Calif., October 23–27, 2004.

The dating experiment is described in M. Snyder, E. D. Tanke, and E. Berscheid, "Social Perception and Interpersonal Behavior: On the Self-Fulfilling Nature of Social Stereotypes," *Journal of Personality and Social Psychology* 35 (1977): 656–66.

Chapter 6: The Power of the Frontal Lobes

Thanks to John Gabrieli for reviewing my description of his research, and for providing me with an advance copy of his coauthored paper, with Kevin N. Ochsner et al., "For Better or for Worse: Neural Systems Supporting the Cognitive Down- and Up-Regulation of Negative Emotion," prior to its publication in *NeuroImage* 23 (2004): 483–99.

The effect of labeling emotions is taken from G. Tabibnia, M. Craske, and M. Lieberman, "Linguistic Processing Helps Attenuate Physiological Reactivity to Aversive Photographs After Repeated Exposure," poster 193.6 presented at the 35th annual meeting of the Society for Neuroscience, Washington, D.C., November 12–16, 2005.

The Mario Beauregard research is taken from "Scientists Unravel Brain Circuits Involved in Joy and Sadness," Society for Neuroscience news release, August 6, 2005.

The reappraisal reference is from Kevin Ochsner, "Characterizing the Functional Architecture of Affect Regulation: Emerging Answers and Outstanding Questions," in John T. Cacioppo, Penny S. Visser, and Cynthia L. Pickett, eds., Social Neuroscience: People Thinking About Thinking People (Cambridge, Mass.: MIT Press, 2006).

Kevin N. Ochsner and James J. Gross discuss the placebo response in their "The Cognitive Control of Emotion," Trends in Cognitive Sciences 9, 5 (2005): 245.

Negative reappraisal is from ibid.

The professional actors study is from E. Perreau-Linck et al., "Serotonin Metabolism During Self-Induced Sadness and Happiness in Professional Actors," program 669.3 presented at the 34th annual meeting of the Society for Neuroscience, San Diego, Calif., October 23–27, 2004.

The Helen Fisher material is from personal communication and Helen Fisher et al., "Motivation and Emotion Systems Associated with Romantic Love Following Rejection: An fMRI Study," program 660.7 presented at the 35th annual meeting of the Society for Neuroscience, Washington, D.C., November 12–16, 2005.

Chapter 7: How Our Brain Determines Our Moral Choices

The Charles Darwin quote appears in his The Expression of the Emotions in Man and Animals (London: John Murray, 1872).

The material on surgeons who speak in dominant tones is from Nalini Ambady et al., "Surgeons' Tone of Voice: A Clue to Malpractice History," Surgery 132, 1 (2002): 5–9.

Nalini Ambady on lie detection is from Y. Susan Choi, Heather M. Gray, and Nalini Ambady, "The Glimpsed World: Unintended Communication and Unintended Perception," in Ran R. Hassin, James S. Uleman, and John A. Bargh, eds., *The New Unconscious* (Oxford, UK: Oxford University Press, 2005).

The Scott Faro study is from "Brain Imaging Could Spot Liars," news@nature.com, November 29, 2004.

Moral dilemmas are discussed in Ralph Adolphs, "Cognitive Neuroscience of Human Social Behavior," *Nature Reviews/Neuroscience* 4 (March 2003): 165–78.

Chapter 8: Make My Memory

The Schmolck material is from personal communication and H. Schmolck, "The Formation of Real Life Memories: Differing Predictors for Facts Versus Events," program 79.12 presented at the 34th annual meeting of the Society for Neuroscience, San Diego, Calif., October 23–27, 2004, and from H. Schmolck, E. A. Buffalo, and L. R. Squire, "Memory Distortions Develop over Time: Recollections of the O. J. Simpson Trial Verdict After 15 and 32 Months," *Psychological Science* 11, 1 (2000): 39–45.

Richard J. McNally discusses people's memories of the O. J. Simpson trial in his book *Remembering Trauma* (Cambridge, Mass.: Belknap Press, 2003).

The Jean Piaget "kidnapping" is described by Andrew Scull in Andrew Scull, "I Remember It Too Well," *Times Literary Supplement*, May 13, 2005.

The Stewart's root beer example is from G. W. Prince, "Yesterday, Today and Tomorrow," *Beverage World*, March 15, 2000.

The other published examples of memory morphing are taken from Kathryn A. Braun, Rhiannon Ellis, and Elizabeth F. Loftus, "Make My Memory: How Advertising Can Change Our Memories of the Past," *Psychology and Marketing* 19, 1 (2002): 1–23.

The Linda Levine study of memory changes after the O. J. Simpson trial is from L. J. Levine et al., "Remembering Past Emotions: The Role of Current Appraisals," *Cognition and Emotion* 15 (2001): 393–417, and is also discussed in Gerald Zaltman, *How Customers Think: Essential Insights into the Mind of the Market* (Boston: Harvard Business School Press, 2003), 181.

The rigged Disney advertisement is described in Braun, Ellis, and Loftus, "Make My Memory."

The Zaltman quote is from *How Customers Think*, 181.

The Coke-Pepsi experiment is from Samuel M. McClure, Jian Li, Damon Tomlin, et al., "Neural Correlates of Behavioral Preference for Culturally Familiar Drinks," *Neuron* 44, 2 (October 14, 2004): 379–87.

The Pribyl research, including the luxury retailer example, is from C. Pribyl et al., "Neural Basis of Brand Addiction: An fMRI Study," program 668.5 presented at the 34th annual meeting of the Society for Neuroscience, San Diego, Calif., October 23–27, 2004.

The Lincoln LS ad was printed in *Smithsonian*, June 2005.

The Linton Weeks quotes appeared in Linton Weeks, "Pod People," *Washington Post*, June 8, 2005, C-1.

The *Science* study is from Uri Hasson et al., "Intersubject Synchronization of Cortical Activity During Natural Vision," *Science*, March 12, 2004, 1634–40.

Chapter 9: Neuroeconomics: What Happens in the Brain When We Reason and Negotiate (*and Why Honesty* Really *Is the Best Policy*)

The Ultimatum Game is described in Greg Miller, "Reflecting on Another's Mind," *Science*, May 13, 2005, 945–47.

The Amy Nelson material is from Amy Nelson et al., "Expected Utility Provides a Model for Choice Behavior and Brain Activation in Humans," program 20.12 presented at the 34th annual meeting of the Society for Neuroscience, San Diego, Calif., October 23–27, 2004.

The Sanfey material is from a discussion at the 12th annual meeting of the Cognitive Neuroscience Society, New York, April 9–12, 2005.

The trust-distrust experiment is from P. Zac, K. Borja, R. Kurzban, and W. Matzner, "The Neurobiology of Trust," program 203.23 presented at the 34th annual meeting of the Society for Neuroscience, San Diego, Calif., October 23–27, 2004.

Oxytocin research is described in "Oxytocin Nasal Spray Makes Test Subjects More Trusting," *Health News*, June 1, 2005.

The revenge-is-sweet experiments are described in Brian Knutson, "Sweet Revenge?" *Science*, August 27, 2004, 1246–47, and Dominique J.-F. de Quervain et al., "The Neural Basis of Altruistic Punishment," *Science*, August 27, 2004, 1254–58.

Chapter 10: The Perils of the Neurosociety

The "office-obsessed employee" is from James E. Challenger, "Career Seeker: To Secure Job, Stay Connected on Vacation," *Richmond Times Dispatch*, June 22, 2005.

The percentage of overworked Americans is from *Time*, July 25, 2005.

The "dying kick of phrenology" is taken from *The Oxford Companion to the Mind*, ed. Richard L. Gregory (New York: Oxford University Press, 1987), 620. The Trevarthen quote is on page 746 of the same volume.

The Zaltman quote is from Gerald Zaltman, *How Customers Think: Essential Insights into the Mind of the Market* (Boston: Harvard Business School Press, 2003), 192–93.

The material on sexual attraction research is from Annalee Newitz, "The Coming Boom," *Wired*, July 2005.

The political applications of neuroscience is from Lisa Phillips, "Brain Scanning Political Views: Is It Valid Science or Marketing?" *Neurology Today* 5, 3 (2005): 29–30.

The Freedman quote is from ibid.

The Heilman quote is from ibid.

The powerful effects of facial similarity are described in Jeremy N. Bailenson et al., "Transformed Facial Similarity as a Political Cue," *Political Psychology*, in press.

Appearance of competence is discussed in Alexander Todorov et al., "Inferences of Competence from Faces Predict Election Outcomes," *Science*, June 10, 2005, 1623–26.

Afterword

The two-gene-variants reference is in Emma Young, "Our Brains They Are A-changing," *New Scientist*, September 17, 2005.

The 24-point IQ increase is from Ronald Kotulak, "Brain Development in Young Children," paper presented at the "Brain Development in Young Children: New Frontiers for Research, Policy and Practice" conference, Chicago, June 13, 1996.

The Ray Kurzweil quote is from his "Human 2.0," *New Scientist*, September 24, 2005.

The Marcus Raichle quote is from *Nature* 435 (May 19, 2005): 254–55.

The Hofstadter quote is from *New Scientist*, October 29, 2005, 52.

Thomas H. Murray's remarks are from his lecture "The Neuroethics of Enhancement," presented at the 35th annual meeting of the Society for Neuroscience, Washington, D.C., November 12–16, 2005.

Distinct brain responses for extroversion are described in B. Haas et al., "Extroversion Predicts Increased Functional Connectivity with the Anterior Cingulate During the Processing of Positively Valenced Verbal Stimuli," program 660.2 presented at the 35th annual meeting of the Society for Neuroscience, Washington, D.C., November 12–16, 2005.

Special Forces findings are from M. Nakic et al., "Emotional Processing in Resilient Special Forces Soldiers," program 660.17 presented at the 35th annual meeting of the Society for Neuroscience, Washington, D.C., November 12–16, 2005.

The Jakob Linnet quote is from "High Sensation–Seeking Men Have Increased Dopamine Release in the Right Putamen During Gambling," program 876.24 presented at the 35th annual meeting of the Society for Neuroscience, Washington, D.C., November 12–16, 2005.

Puppies monitoring human gaze is from Sarah-Jayne Blakemore, Joel Winston, and Uta Frith, "Social Cognitive Neuroscience: Where Are We Heading?" *Trends in Cognitive Sciences* 8, 5 (2004): 216–22.

The Ray Kurzweil quote is from "Human 2.0."

Obese women and striatal activation is from Y. Rothemund et al., "Caloric Content of Visually Presented Food Items Modulates Cortical and Subcortical Activation in Obese Women—an fMRI Study," program 529.6 presented at the 35th annual meeting of the Society for Neuroscience, Washington, D.C., November 12–16, 2005.

Index

About the Author

RICHARD RESTAK, M.D., a neurologist and neuropsychiatrist, is the author of seventeen previous books on the human brain, including the bestsellers *The Brain* and *Mozart's Brain and the Fighter Pilot*. His essays have appeared in the *Washington Post*, *The New York Times Book Review*, the *Los Angeles Times*, and the *American Scholar*. A Clinical Professor of Neurology at George Washington University School of Medicine and Health Sciences, he serves as president of the American Neuropsychiatric Association.

Also by RICHARD RESTAK

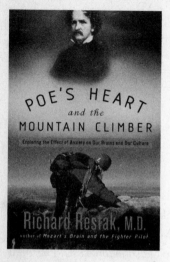

POE'S HEART AND THE MOUNTAIN CLIMBER
Exploring the Effect of Anxiety on Our Brains and Our Culture
$12.00 ($17.00 Canada)
978-1-4000-4851-9

MOZART'S BRAIN AND THE FIGHTER PILOT
Unleashing Your Brain's Potential
$13.95 ($18.00 Canada)
978-0-609-81005-7

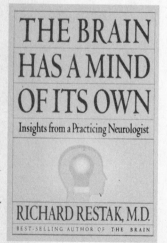

THE BRAIN HAS A MIND OF ITS OWN
Insights from a Practicing Neurologist
$15.00 ($22.00 Canada)
978-0-517-88080-7